THE RELUCTANT FARMER

THE RELUCTANT FARMER

Memoir of an Unexpected Journey

CATHRYN WELLNER

Copyright © Cathryn Wellner, 2025
25/13 South Esplanade
Glenelg, South Australia 5045

All rights reserved.

No part of this book may be reproduced in any form or by any electronic or mechanical means, including information storage and retrieval systems, without written permission from the author, except for the use of brief quotations in a book review.

Cover photographs © Richard Thomas Wright

ISBN 978-1-7642854-0-7

 A catalogue record for this book is available from the National Library of Australia

CONTENTS

Preface ix

Part One
AUCHINACHIE FARM, VANCOUVER ISLAND, 1990-1994

1. Facing the Red Lions 3
2. Learning to Be Canadian 8
3. We Bought the Farm 12
4. Best Wedding Ever 14
5. The Thin Edge of the Animal Wedge 17
6. Bird Brains & Rotten Eggs 20
7. When Cocksure Became Cock Caught 22
8. Pick on Someone Your Own Size 24
9. Chicks & Farming Lessons 26
10. But We Will Have to Kill Them 29
11. Free Range Guilt 32
12. Mama Can't Outwit Crows 35
13. The Farmer & the Egg Thief 38
14. Heroism in a Small, Fierce Package 40
15. Trust from a Wounded Chicken 43
16. Chick on the Loose 45
17. Turkey Liberation Front 47
18. So That's Why They Clip Their Wings 51
19. Brown Thumb Gardener 53
20. Interfering with True Love 56
21. The Orphan's Cry 59
22. Doesn't Everyone Have a Pet Bat? 63
23. Fred Eaglesmith & Farm Heartaches 65
24. A Corker of a Porker 68
25. A Pig Lover & Goddess Worshippers 71
26. A Sweet Way to Go Broke 74
27. The Sheep Midwife 77
28. Too Much of a Good Thing 80
29. Auction Is a Seven-Letter Word 83

30. Beware the Battering Ram	85
31. Not Quite Enough Courage	87
32. Animals Do Talk on Christmas Eve	89
33. Lessons from the Sheep Barn	93
34. Ovine Escape Artist	96
35. "If You've Got Livestock, You've Got Dead Stock"	98
36. Sprinkle with Hypocrisy, & Serve Hot	101
37. Mothering the Weak Ones	103
38. Island Interlude	106
39. A Dream Strong As the Lure of Gold	109
40. A Daunting Introduction	112
41. Ma & Pa Head North	114

Part Two
PIONEER RANCH, CARIBOO, 1994-2004

42. Arriving at the Ranch	123
43. Have You Met the Weird Neighbours?	126
44. Cathryn's Cariboo Kitchen	129
45. The Pioneer Ranch Restaurant	131
46. Fire in the Woods	134
47. Searching For Pigs	136
48. Celebrating Cariboo	138
49. Cementing My Outsider Status	140
50. The Height of Optimism	144
51. Truck Driver to the Rescue	147
52. And the Cat Came Back	149
53. Cold Comfort	151
54. Rescuer Ram	154
55. The Grief of a Grey Goose	156
56. The Indomitable Red Hen	159
57. Wild Goose Summer	162
58. Death Visits the Ranch	165
59. The Deep Soul of a Dog	168
60. Phantom Pays the Price	170
61. Best Friends Until They Are Enemies	174
62. "Do You Want Her Back, or Shall I Shoot Her?"	177
63. "I Didn't Know You Were Educated"	180
64. Bearing Witness Is a Gift	185
65. Wild Neighbours, Wild Life	187

66. Bears in My Back Yard	189
67. Cheering for the Bull	191
68. Angels in the Operating Room	196
69. "Hey, I'm an Indian"	199
70. Oh, Where Has Our Little Dog Gone?	201
71. Hoping For Miracles	204
72. Eagerly Planting Weeds	206
73. Pig Tusks and Close Calls	210
74. Death of Black Boy	213
75. Good Fences Make Good Neighbours	215
76. Outrageous Courage	217
77. And Then There Were Camels	220
78. Making a Camel Feel at Home	222
79. When a Camel Gets Urges	225
80. An Unplanned Rodeo	228
81. High Hopes in Wells	232
82. A Sad Farewell	235
83. Standing Broad Jumps	240
84. Finding My Sealskin	243
Epilogue	250
Acknowledgments	253
About the Author	255
Also by Cathryn Wellner	257

PREFACE

Country living held little allure for me, though I enjoyed quiet retreats in peaceful, rural settings. So when my friend and fellow storyteller, Cherie Trebon, invited me to the Spotted Chicken Reunion, I was intrigued. The annual gathering was the brainchild of Jeanne Hardy, who published *The Spotted Chicken Report* from her rural acreage in Twisp, Washington. For Jeanne, a "spotted chicken" was a chicken with spots—or a chicken that had simply been spotted. Either was reason enough for a good laugh and a celebration of country life.

Cherie and I drove over the mountain pass, up the Loup Loup Highway, and into Twisp. A side road led us to a barn-shaped log cabin on a wide, grassy acreage. There we met a laid-back, fun-loving group that included *Spotted Chicken* fans, Jeanne's friends, and others who shared her delight in the comic possibilities of rural living.

That weekend introduced me to a woman who would become a dear friend for the rest of her life. A few years later, I found myself transformed from a city dweller into a country newcomer—an American making a home in a land that viewed its southern neighbour with both affection and wariness. My sudden move from Seattle to British Columbia surprised my friends, and some quietly drifted away. Uprooted and uncertain, I had much to learn about country life.

Jeanne understood. Once a city woman herself, she had moved her four sons to rural Washington and found both joy and challenge in her new world. She understood my heart, my history, and my dreams. She knew the shift from city to country could be disorienting but also rich with unexpected blessings. I could write anything to her. She took it all in with a wide-open heart. With her, I was safe. From her, I could learn.

Our correspondence deepened over the years. She listened, laughed, and encouraged. When things became untenable, she gently urged me to act. When I didn't, she stayed with me but reminded me who was in charge of my life.

When she died on November 13, 2002, I was devastated. I had lost my anchor and beloved friend. We would no longer share the laughter and challenges of her life in Twisp or my unfolding adventures in California, British Columbia, and Australia.

As I revisit those surprising years, I can almost hear her responses—her laughter, her tears, her gentle admonitions. These chapters unfold in loose clusters of experience rather than strict chronology, much like the letters we once exchanged.

This one is for you, Jeanne.

A note on spelling and punctuation:

Writers in the U.S., Canada, and Australia follow slightly different rules. I learned in the American education system, wrote professionally in Canada, and now write in Australia. Generally, I follow the rules of my early years, but after thirty-plus years in Canada, I add u's to words like *neighbour* or *colour*. The "er" of my earlier years becomes *metre* or *theatre*. I make mistakes and am inconsistent. I blame my old brain, but I don't think the errors will interfere with the story itself.

Part One

AUCHINACHIE FARM, VANCOUVER ISLAND, 1990-1994

Chapter One
FACING THE RED LIONS

"How would you like to do me a favour and take my place at a conference in Vancouver?"

Jimmy Neil Smith's Tennessee drawl was unmistakable. So was his dislike of travel. Since I was living in Seattle, Washington, he figured it would be an easy trip for me.

Years earlier, he had founded the National Association for the Preservation and Perpetuation of Storytelling. The ancient art was undergoing a major revival and morphing into something more varied than performance. People in business, politics, education, non-profits, law, and health were learning to use stories to communicate more effectively. A major tourism conference had invited Jimmy Neil to talk about using storytelling in the travel industry. I was serving on the NAPPS board when he called and was eager to share my enthusiasm for storytelling with a Canadian audience.

The participants who came to hear about storytelling hung on my words and seemed to grasp how storytelling could enhance their marketing and offer a better experience for their customers. One participant was particularly enthusiastic. He and his wife owned Pine Lodge Farm on Vancouver Island. The businessman understood my message, but his real dream was to take the stage as a storyteller. Before the conference ended, he asked if I would be willing to travel to the island. He wanted to organize an event in his B&B, featuring him, me, and the editor of one of the two local newspapers.

A snowstorm buried the first planned date. Two months later, the weather cooperated. A long drive from Seattle took me across the border and to the ferry terminal at Tsawwassen. From there, I caught the ferry to Sidney on Vancouver Island and drove up the Island Highway to Mill Bay. The scenery was spectacular. The large, wood-lined lounge of Pine Lodge Farm was an inviting venue. The owner and his wife welcomed me warmly. The performance—for a packed house—was a hit. The owner had a repertoire of humorous tales. Richard, the editor, proved to be a gifted writer. I was in my element. The three of us were flying high, buoyed by the enthusiastic listeners.

Richard offered to take me for a drive the next day. We felt the electricity between us crackle. Weeks later, he came to the city. The connection was still there, and we began making plans. I would try out life on Vancouver Island. If I did not like it, we would move back to Seattle. The idea was extravagantly impetuous for someone with a modest, careful bent. A few years earlier, I had resisted my first husband's suggestion that he accept a position at the University of British Columbia. I was a contented city woman. I swore I would never leave Seattle, where I was known for my storytelling.

That was the first of many "nevers" I would let go of in the following years.

Richard was renting space for his mobile home on land surrounded by farm fields. He had built a deck off the front door. On soft evenings, we could sit and watch the neighbour's cows watching us and catch the scent of the flowers he had planted in the garden. Beside it stood a small stand of tall fir trees. On clear nights, the stars were brighter than anything I had seen in years. Country life was not part of any life I envisioned, but this was more like country light. We were on the outskirts of Duncan, one of a cluster of small towns spread along the Island Highway. It was only an hour's drive to Victoria, British Columbia's capital city, and a ferry ride from Seattle or Vancouver.

Nature had always been a major draw for me, nurtured by my small-town upbringing and the evocative writing of two of my favourite writers. Kathleen Norris's *Dakota* and Gretel Ehrlich's *The Solace of Open Spaces* sometimes made me yearn for a quieter setting. Now I would have it.

Knowing the transition from city to country, from large house to small mobile home, would be a major adjustment, Richard turned the addition (an old miner's cabin attached to the mobile home) into my office. While he made do with a tiny room packed with books, he gave me a large space that looked out on the green fields. It was my retreat, where I asked myself repeatedly, "What on earth have I done?"

I could relate to Lot's wife. Preachers chide her as disobedient, but I saw the story from her perspective. She was leaving behind two of her daughters and the familiar comforts of the city to follow her husband into the mountains. She knew they were moving from a nice home to a tent. Sodom and Gomorrah were wicked cities, but that is where her friends and family lived.

As they fled from everything she knew and loved, she turned around for one last look at what she was leaving. Some pretty uncharitable angels had warned them not to look back. For ignoring the warning, Lot's wife was turned into a pillar of salt. That sounds as mean-spirited as destroying the cities' innocents along with the bad actors.

Had the Big Guy turned me into a pillar of salt for all my looking back—at the city, the country, the friends, and the work I had left—I could have flavoured thousands of dishes at thousands of feasts. The summer before I moved to Vancouver Island had been a time of questions, fears, and doubts. Friends' reactions were mixed. Some were open and supportive. Others backed away.

Norris Spencer remained stalwart in her support. She encouraged me to talk through all the confusion. I never had to hide with her, nor pretend more certainty than I felt.

Understanding how important Norris was to my making peace with myself, Richard gave me a special birthday gift. He offered to pay for her ferry to the Island. She came in November and was there for our first community storytelling evening. She laughed and sang with us for hours and accepted the changes in my life with an openness that was soul healing.

We took her to hangouts that had become our favourite places. At the Arbutus Café for Saturday breakfast, we chatted with writers, artists, and cultural creatives. She joined us for a curry feast with friends and an evening at the Folk Guild. We stopped by Pine Lodge Farm, where we had met, and stayed for pizza. Then we came back home and sang for hours. Norris and Richard hit it off. They formed a bridge for me, between the old and the new.

My storytelling work was still south of the border. For years, I had taught storytelling to students in the Master of Education program at Pacific Lutheran University. That summer, I taught a Persian tale I had learned from Diane Wolkstein. In it, a young man flees a red lion, only to find red lions at every turn until he stops running and faces his fears. When he finally confronts the red lion, he discovers it is tame. Only fear makes the lion dangerous.

A woman in the class was fleeing her own red lion. She wrote to me afterward:

"I related so strongly to the Red Lion story that the mere mention of it made tears well up... I realize I need to have a direction, take a stand, and not fear so much the consequences. It's time to deal with the Red Lion."

Her letter turned a light bulb on in my brain. I wrote her a reassuring letter, gave Lot's wife a compassionate hug, and turned to face my red lions.

They were pretty tame. Mostly, they were just the usual stumbling blocks accompanying any major move, made more complicated when it entails changing countries. I was still a champion worrier, as had always been my nature. Sometimes I felt like one of those cartoon characters who are running so fast they do not know they have gone over a cliff. Their arms and legs keep pumping as they plummet to the ground.

Before I could self-destruct, I began to sprout wings. They were wrinkled, wet, and weak, but they were wings. As they grew, dried, and spread, they not only kept me from crashing. They held me aloft during training flights. They grew strong and sure.

Richard was well connected in the community. His friends tossed me lifelines and gradually pulled me out of my stuck place. The writing assignments he sent my way introduced me to other people in our small town.

By November of 1990, I was feeling more at home in my new life. Richard encouraged my writing and proved to be the kind of editor every writer should have. He knew how to unearth my buried ledes and make a flat piece edgy, without changing the intent or making it sound like someone else had written it. In turn, I encouraged his storytelling, nudging him to lift his carefully crafted stories from the page so he could keep eye contact with the audience. We began performing together regularly. We both sang, and he played autoharp. We wove music through our stories and created programs that worked well with two voices, two perspectives.

The image of Lot's wife began to fade, or at least of her turning into a pillar of salt. She never had a chance to embrace a new life. I did, and it was about to take me in surprising new directions that began with the question, "You want to look at a farm?"

But first, I needed to absorb some lessons about becoming a Canadian. I had lived a year each in France, Germany, and the Netherlands. Because the stays were short, I had the luxury of being an observer. This time, I wanted to learn how to be at home.

Chapter Two
LEARNING TO BE CANADIAN

Hockey, snow, and polite people. That was my image of Canada.

Like many Americans, I knew little about the vast country to the north. It was simply there, reliable and mild-mannered—a vanilla ice cream kind of place, probably made of quality cream but without a distinct flavour. I hadn't learned enough to form real misconceptions and just viewed it as a friendly neighbour.

The adjustment was more difficult than I expected. I assumed Canadians were like us, only colder. So I was caught off guard when offhand comments about Americans sometimes revealed an undercurrent of resentment. In Seattle, I had surrounded myself with progressive, forward-thinking people. My friends and family did not match the stereotypes often attributed to Americans. And yet, I arrived with an embedded belief that I came from the best country in the world—a belief I hadn't fully examined until I heard it reflected back to me with skepticism.

I carried my myths across the border, right alongside my boxes and books. I had devoured the stories of Davy Crockett, Daniel Boone, and George Washington along with my childhood breakfast cereal. Canadians, I discovered, told different stories. The War of

1812 had a contrasting outcome on the northern side of the border, and Queen Elizabeth was more than a figurehead. She was part of the constitutional fabric. The people who lived there before settlers arrived were First Nations rather than Native Americans. The province was headed by a premier, not a governor. The country had a prime minister, not a president.

Everyday objects had different names. Our sofa was a chesterfield. The wool caps we pulled on when the temperature plunged were toques. The *ah* sound in pasta and garage became flat *a's* that sounded harsh to my ears. When I wrote for local publications, I had to become a bilingual speller, adding u's to color and neighbor. We shared a common language but not always the same worldview.

Teaching my first storytelling classes on Vancouver Island, I discovered how U.S.-centric my outlook and images were. I knew how to be an American. I had to learn how to be a Canadian. In many ways, the transition was more difficult than learning to thrive in France during my year of graduate study there. I expected the language and culture to be different in France. I expected them to be familiar in Canada. They were—and they weren't.

I brought my red-white-and-blue identity to my new home. I had grown up with the story of the melting pot—the idea that anyone who came to America had been fortunate to find refuge in a land of opportunity, and that newcomers were encouraged to blend into a shared national identity. In contrast, Canadians often spoke of a mosaic, where immigrants enriched the cultural fabric by retaining their traditions and perspectives as they settled into Canadian life.

That mosaic ideal, however, did not always extend to my country of origin. While Canadians generally celebrated the diversity of new immigrants, they often assumed Americans would adapt easily and quickly. I might have been less surprised by that had I grown up with a deeper understanding of the land to the north, largely missing from our school lessons. Some of my Pacific Northwest friends dreamed of creating a bioregion uniting Washington,

Oregon, California, and British Columbia. Yet in three decades in Canada, I never met anyone who saw "Cascadia" as a welcome idea.

The transition from the U.S. to Canada might have been easier if the distance had been greater. My old home was only a few hours away from the new one, but the border created an illusion of separation far more profound than the map suggested. Even the ferry ride that punctuated the drive made the crossing feel symbolic—a passage from one way of belonging to another.

Until my status in Canada could be sorted out, I continued to work on the American side of the border. Canadian officials always asked if I was carrying firearms. Once I started the paperwork to become a landed immigrant, I had to trade my Washington State licence plates for British Columbia plates. Same car, same woman, but I was never again asked about guns.

On the other hand, I had to start proving to American customs officers that I had a right to work in the States. This was before passports were required for entry into the U.S. I would arrive at the border and flash my British Columbia driver's licence as identification. When the customs officials asked the purpose of my visit to the U.S., I would smile innocently and say, "I am going for work."

Their faces would switch into alarm mode as they launched into a spiel about my needing permission to work in the States. I would say, "Oh, sorry," as if I had just realised my faux pas. Then I would reach into my purse and pull out my American passport. That was in the days before the attack of 9/11 and the election of a certain narcissist. Border crossings were mostly simple and friendly, and a certain amount of banter with customs officials was safe.

During my second summer on Vancouver Island, I was reminded how little many Americans know about Canada. We had moved to a small farm, and with its wide-open space and beautiful setting, I decided to host a storytelling retreat. Among the participants was a teacher from Washington State who would soon be teaching a fifth-grade class with a unit on Canada.

One evening, as our group gathered around the dinner table,

enjoying the easy camaraderie that had developed, the teacher admitted something with a laugh: she had been surprised by the warm weather. She had imagined that crossing the border into Canada meant entering a land of snow and ice—a place both colder and somehow caught in an earlier time.

Our farm, below the 49th parallel, was basking in a July sunset. The Canadians around the table chuckled good-naturedly at her surprise and cheerfully answered her questions.

In that moment, I realised I felt Canadian for the first time. I had absorbed more than I'd realised and had begun to feel at home. The folk music, the rhythm of small farm life, the new friends, and even the different places to shop felt familiar now. My sense of self was returning from its long wandering in the desert of displacement. I was becoming like a turtle, learning to carry home on my back and to embrace a certain rootlessness. Though I sometimes envied those who stayed in one place for decades, I was grateful to understand that home could be wherever there were people to love, and to love me in return.

My connection to Canada deepened when Mother told me that some of her siblings had been born there. Her parents had fallen on hard times in Nebraska and moved north, drawn by the promise of cheap land and a fresh start. They farmed no more successfully in Canada than they had in the States and eventually returned south after two of their eleven children were born.

Later, Mother added that one of her maternal great-great-grandmothers had been among the first French women to come to Canada. The link between that family lore and the Nebraska homestead faded into uncertainty, but the possibility of Canadian ancestry gave me a small, unexpected sense of belonging.

The longer I lived in Canada, the more I found it suited me. Its emphasis on community, multiculturalism, and universal health care echoed the values that had shaped me from childhood.

As it turned out, though, learning how to be a Canadian was a gentle adjustment compared to what was coming next.

Chapter Three
WE BOUGHT THE FARM

Real estate prices in our island town were still low enough for us to dream within reach. We could afford a modest but comfortable home in a middle-class neighbourhood—one I could buy outright without touching my savings or retirement accounts. I pictured a simple house with no mortgage and little yard work, a place easy to lock and leave when work or wanderlust called. I threw myself into the search.

Richard had never spoken of longing for land, but he was open to possibilities. One evening, as we wandered through a home show, we stopped at a real estate booth. An ad caught his eye, and we decided to drive by the property. From the road, the house seemed too large and too close to the street, but curiosity nudged us to call the realtor and take a closer look.

The small farm, built around the turn of the twentieth century, offered eleven acres for dreaming. We fell in love almost instantly—with the sturdy four-bedroom house wrapped in a verandah, the wide views across green fields, and the twin humps of Mount Prevost rising in the distance. A one-bedroom rental on the property promised enough income to cover a small mortgage, and rent from Richard's mobile home would help even more.

Beyond the house stood a big blue barn with a peaked roof. The ground floor held stalls; the open loft above seemed made for storytelling evenings and music gatherings. We began imagining workshops, laughter spilling into the rafters, music echoing across the fields.

There would be room for a generous vegetable and herb garden, for flowers and a small flock of hens. Green fields would buffer us from the slow creep of development. Nearby dairy farmers might rent the pasture. A greenhouse, a few sheds, and mature fruit trees felt like icing on the farm cake.

It was a larger dream than my tidy, mortgage-free vision, but the lure of space and beauty was irresistible. We signed the papers and moved in.

Upstairs, we each claimed an office on the mountain side of the house; the master and guest bedrooms faced the street. We dreamed of one day converting the high-ceilinged loft into our shared workspace and turning the bedrooms into a bed and breakfast.

Life soon settled into a gentle rhythm. Each morning, Richard headed off to his job as editor of a local newspaper. I stayed home, sending out publicity, polishing new stories, and writing. We had space, light, and big dreams.

For a while, the rent from the mobile helped. Then a tenant vanished, leaving it in shambles. The small-claims judgment in our favour proved worthless; we never found him. We cleaned, repaired, and sold the place, rolling the modest profit back into the farm.

We told ourselves it was just a small setback—the kind any dream demands. But sometimes dreams grow heavier than we expect, their beauty concealing the weight we'll eventually carry.

Chapter Four
BEST WEDDING EVER

Divorce had convinced me I was a faulty person. My new partner was willing to wait while I came around to the idea that a second marriage would not be absolute proof of that. In the meantime, I travelled to the U.S. to perform and give workshops. I could remain in Canada for six months without a visa, but I could not work.

We had begun performing a combination of music and storytelling. Audiences seemed to particularly enjoy one of our programs, *Partners for the Long Dance*, a wry, poignant look at relationships through stories and songs. We often included *Dear Old Buffalo Boy*, sung back and forth by a man and a woman planning a wedding—until she discovers he already has six children.

The woman asks, "When will we be married?"

He replies, "I guess we'll get married in a week."

One night at the folk guild, Richard changed the words. When I sang, "When will we be married," he didn't miss a beat.

"How about June 29th?" he said.

The audience had been watching us for months. No one laughed, but we could see speculation on their faces. Richard filled a long pause with chords on his autoharp, then we picked up the song and went on.

When we got home, we looked at the calendar. June 29, 1991, was a Saturday. We had no other plans, so we called the bed and breakfast where we had met and booked the whole place for a wedding.

Neither of us had any religious affiliation, but Richard knew of a new minister in town. He had asked me to interview her for the local paper. I liked her from the first question—warm, open, and not in the least dogmatic. She seemed just the person to tie the knot for two retreads. She readily agreed.

We began compiling a list and sending out invitations. In my year on the island, I had been welcomed by Richard's friends, so we included them, along with family and friends from both sides.

Our guest list reflected our creative lives: writers, storytellers, musicians, poets, photographers, and artists. As acceptances flowed in, we were convinced we were planning the best wedding ever. The rooms in the lodge bed and breakfast filled quickly; nearby motels accommodated the overflow. Guests came from as far east as New York, as far south as California, and as close as our own town.

Unlike many newlyweds, we did not need pots, pans, or towels. Instead, we asked guests to bring their talents. On our wedding day, we looked around at the smiling faces and felt the warm embrace of our scattered community.

The wedding nearly happened without me. Vi Hilbert, a beloved Upper Skagit elder and long-time friend, gave a blessing in Lushootseed and English. The minister welcomed the gathering and then began the ceremony. I was still upstairs, waiting for the signal to make my grand entrance. The startled groom stopped proceedings and sent his youngest son to fetch the bride.

When the ceremony ended, the party began. Richard's oldest son shone as master of ceremonies. Our friends and family gave speeches, shared stories, songs, and poems. It was the best wedding we had ever attended—and it was ours.

We hosted a pancake breakfast the next morning, inviting family and our out-of-town guests. Auchinachie Farm was already becoming a gathering place where all were welcome. Storytelling and music would soon ring from the rafters and across the fields.

After the last hugs, the final happy tears, and the warm goodbyes, we settled in to figure out what we would do next. Life seemed full of promise.

Chapter Five
THE THIN EDGE OF THE ANIMAL WEDGE

Before the wedding, Richard's oldest son gave us a gift: a black-and-white sheep-herding dog he thought a farm needed. The Border Collie had a mask like Erik in *Phantom of the Opera*, so we named him Phantom. He was adorable and loving—but thick as a post. He must have been hiding when smart genes were being passed out to his litter.

My years of being free to shut the door and travel, without worrying about who would look after pets, ended abruptly. Phantom needed feed, water, and a lot of training. Since Richard was working full time, the task fell to me. I proved abysmal at it.

Dog training was not in my skill set, so we found an instructor. Classes were held in the multi-purpose room of a school. Phantom and I failed the first lesson, circling the room. Phantom wanted to meet every other candidate and challenge half of them. I tried, unsuccessfully, to rein in his exuberance. Then there was the heating system, which blasted air through a large grate on the front side of the stage. Whenever we neared it, Phantom flattened his body, extended his legs, and refused to walk. Red-faced, I dragged him like a dust mop. Once past the threatening grate, he leaped to his feet and continued happily disrupting the attention of the other participating dogs.

We lasted two sessions. After the second one, the trainer took me aside and said, "Don't come back until you have better control of your dog." Control? I had none. That was why I had enrolled in the course.

We slunk to the car and drove home. I could not afford private lessons. Phantom would have to learn by trial (his) and error (mine). He never retained much beyond his one trick. Given the promise of a treat, he would sit, roll over, and shake a paw.

We almost did not reach the training stage because of the pup's friendliness and sense of adventure. Somehow, we had missed the message that dogs must be licensed at four months. He had barely passed that milestone when he disappeared. We searched the neighbourhood, called everyone we knew, knocked on doors, and plastered the area with posters of our missing dog. And we wept. The independent rascal had clamped his needle teeth on our hearts.

First thing the next morning, we drove to the animal shelter, arriving minutes after someone had dropped Phantom off. "Is he a friendly, well-behaved dog?" asked the staff member. Friendly? Definitely. Well-behaved? Not so much.

He had wandered down the street on the first of many runaway adventures and ended up at a motel. A businessman in town for the night took him in and dropped him off at the shelter when it opened. The two minutes between the businessman's arrival and ours cost us sixty dollars. Some of that was the fine for having an unlicensed dog; the rest was for the licence itself.

Local songwriters had written a song about a little black dog who ran away, wanting to see the northern lights and the sea. One line always brought me to tears: "Little black dog, please come back to me." We added it to our repertoire, but I could never get through it without a lump in my throat. It was more than a song about a beloved dog. It was a song about me. I had answered the call of adventure. Some friends back in the States would probably always see me as a runaway, but this new life was full of enriching experiences.

The little black dog was the thin edge of the animal wedge. I knew nothing about training a dog—and even less about the surprises that lay ahead.

Chapter Six
BIRD BRAINS & ROTTEN EGGS

The local auction was a source of weekly entertainment. A fair bit of that entertainment found its way back to Auchinachie Farm. Never having been to an auction before, I was unprepared for how seductive it could be.

Chickens, ducks, geese, lambs, tools, and trinkets gradually made the trek from sale barn to farmyard, populating our small acreage with dizzying speed. The diversity of our purchases delighted us, though I constantly worried about the unexpected expenditures. The chickens were my favourites, likely because of their daily gift of eggs.

For the first time, I began to understand the etymology of some old idioms. "Clip their wings" was one of them. Although we refused to do it, we understood why farmers would opt for that. The shyer chickens were quite happy to retreat to the safety of the coop when darkness fell. The gutsier ones "flew the coop," spending the night in the cedar hedge that lined our driveway.

The flock quickly established its "pecking order". Those on the low end were regularly "hen pecked." How they determined who ranked highest mystified me, but those in the upper echelons were like schoolyard bullies. They were the first to devour food scraps they saw as special treats. The low-status chickens lost feathers and risked serious injury if they dared challenge the dominant fowl.

Egg gathering was one of my favourite chores. The shy hens dropped theirs in the coop's nesting boxes, while the flyers stashed theirs beneath blackberry bushes, in grassy patches, behind sheds, or anywhere else they felt safe. Eggs were a good source of income. I hunted for them daily and learned the importance of not putting "all my eggs in one basket." Some of those lovely spheres eluded me until I inadvertently stepped on them. If they had been hidden long enough, they confirmed the cattiness of a childhood taunt, "The last one in is a rotten egg."

The broody hens, happily sitting on any egg, whether it was theirs or not, were my favourites. They were the real "mother hens," attentive and caring with their chicks. It seemed like a miracle, watching tiny, wet beings peck their way out of eggs and into the "feathered nests" their mothers lined for them. The hens crooned softly, a sound the little ones recognized from their time in the shell. When the chicks wanted comfort, safety, or warmth, they would run to their mother, who would "take them under her wings" in a way that made yet another saying crystal clear.

We learned not to "count our chickens before they hatched" since not every embryo survived. The hens knew when they had a "bad egg" in the clutch and would roll it out of the nest. Eventually, they were no longer "spring chickens," but the old hens were still good company. We let them hang around the "hen house" until their days ended.

Far from being "bird brained," the chickens were smart and funny. Even those who preferred to fly into the hedge at night knew to "come home to roost". They just had a different concept of home than we did. We enjoyed the sociable creatures and did our best to learn from them—and occasionally got schooled.

It was a big, brassy rooster who taught us the meaning of "cock of the walk."

Chapter Seven
WHEN COCKSURE BECAME COCK CAUGHT

The neighbours raised bantam roosters. One of them became a regular visitor to our farm. He preferred the pulchritude of our feathered ladies to the bachelor pad back home. As long as he steered clear of our roosters, he got away with it, and we left him to wander freely.

One night I took Phantom out to answer his last call of nature. That entailed a final check of the barn to ensure the sheep were all tucked in. Ours was a suburban farm. Occasional packs of marauding dogs made it unsafe to leave sheep outside overnight.

Our animals were fine that night. The neighbour's rooster was not. He had found his way into the room where we stored unsold wool from the previous year's shearing. Wool covered his legs like chaps. In his attempts to get clear of it, he had spun a heavy, unbreakable tie that bound his legs to each other.

He could waddle. He could fly well enough to keep out of reach. And, oh, could he scream. The latter was not unusual. The rooster screamed regularly to assert dominance among the handful of our hens who followed him. This time, however, he outdid himself. He shrieked and flew and paced and ran, screaming as if I were trying to kill him. The barn was in an uproar. The sheep paced restlessly. With no hope of catching the rooster, I turned my attention to a pregnant ewe. She was still still enormous and too relaxed to be near giving birth.

While I stood watching her, a Southdown lamb pushed his head against my left leg. His mother approached and rubbed her head against my right leg. Then a ram we called Old Papa trotted up, head lowered, and gently pushed against my left leg. Not to be left out, the other Southdown lamb squeezed in between his twin and Old Papa. They stood patiently until I gave each a scratch behind the ear, on the haunch, under the neck, or wherever their favourite spot happened to be.

Phantom was beside himself with jealousy and began barking frantically. When I wandered over to calm him, I heard a thumping sound. One of the youngest lambs was springing around the barn, leaping and kicking, glorying in his woolly body.

By this time, the rooster had forgotten about me. I slowly crept up on him and snatched him by the legs. Knowing if I held him upright, he would flail me with his wings, I held him upside down. As soon as he was still, I gently untangled the wool that wrapped his legs.

His screams split the night air, but I held on tightly. The sheep stared in frozen fascination, unsettled by the screaming rooster dangling from my hands.

After a few minutes of tugging and unwinding, I freed the rooster's legs. When I set him down on a railing, he screeched one last time, fluffed his feathers, and strutted away.

Had he been a tad less cocksure, he might have lived out the remainder of his life wandering the farm. Unfortunately, he went in search of adventure. He got more than he bargained for.

Chapter Eight
PICK ON SOMEONE YOUR OWN SIZE

We fell in love with a gorgeously coloured auction rooster. We had no clue what breed he was until some hens started laying blue and green eggs. We were as astonished as our customers were—until we learned that our handsome rooster was an Araucana and that we had somehow brought home some hens of the same breed. They were a large breed that originated in Chile. Their eggs were a visual treat. We knew the colour of the shell bore no relationship to the egg's nutritional value, but our customers expected brown shells.

We sold eggs on the honour system, putting the cartons in an outdoor refrigerator, with a can for payment. That sounds like a recipe for going broke, but most people were honest.

The only problem was that customers thought the blue and green eggs were defective. They would take out a carton, see the colourful eggs, look to see if we were watching, then switch the blue and green ones into a different carton.

For a few days, we made a point of being there when people stopped for eggs. "Did you get any of the special, coloured ones?" we asked cheerfully, telling them about the Araucana and implying the eggs were exceedingly rare. The idea backfired. Word spread that we were selling exotic eggs. Some of our customers started surreptitiously switching out all the brown or white eggs, replacing them with coloured ones.

The Aracauna did more than make our eggs more artsy. He earned his keep by protecting the hens from predators—including the neighbour's cocky bantam.

Normally, the chicken yard was peaceful. Then one day, it sounded as if a fox had invaded. We came on the run and found our Araucana in the process of slashing the cocky bantam to bits. We rushed in, grabbed the little rooster, staunched his bleeding, cleaned him up, and delivered the battered bird to his owner.

Things were quiet for weeks. The Araucana had established his dominance. The hens were at peace. The bantam stayed home.

Alas, the lure of those fine hens was too much for the bantam. When he recovered enough to fly, he returned to our chook yard and challenged the Araucana. He was probably a quarter the size of our rooster, but he had attitude.

That was not enough to protect him. He lost the contest. It was a sad walk, from our farm to the neighbour's door. It was also the last time any of the show birds came after our auction rooster or his harem of lovely hens.

On the other hand, our customers continued to be served with a mix of colourful eggs. One day, we overheard a couple muttering over the price of them. They were clearly incensed that we were charging so much, though they were always fairly priced. They were about to put the eggs back in the refrigerator when we heard another voice. This time it was Longevity John Falkner, a friend from the folk music community. He gave them a brief but pointed response about the value of fresh, free-range, farm eggs. They never complained again.

Chapter Nine
CHICKS & FARMING LESSONS

The soft white feathers and fuzzy head captured our hearts. We bought the Silkie hen at the auction and gave her free run of the farm. We had read that Silkies benefit gardens because they eat bugs and weeds but do not munch the vegetables.

Unfortunately, the gardening season was months away. Watching her wander around the farm, we could sense she was a broody hen. As chickens with a motherhood mission, they want to lay a clutch of eggs and settle in to hatch them.

We assumed her eggs would be infertile. We had no Silkie rooster and had never witnessed her being wooed by one of our male chickens. We did, however, have a Rhode Island Red rooster and a Barred Rock rooster who had the same freedom of the farm. Their courting was discreetly out of our vision.

The Silkie began laying eggs regularly and guarding them fiercely. She gathered her future brood under her wings and concentrated on hatching them. We shook our heads and wondered when she would realize she was sitting in vain. To save her the disappointment, I stole an egg from a hen I knew had been with a rooster and slipped it under the Silkie's soft feathers. She clucked her disapproval, then settled her small frame over the newcomer. So I stole more eggs for her.

She had been sitting on seven eggs for two weeks when a more knowledgeable friend told us we were going about it the wrong way. We should set aside a half dozen eggs, put them in the incubator, turn them over several times a day, and then place them all at once under the broody hen. Otherwise, she would abandon the other eggs after the first one had hatched. In hindsight, that made no sense. Hens have always laid eggs one at a time and looked after the new chicks until the other viable ones pecked their way into the world.

With the friend's advice in mind, I started over. Over the next week, I faithfully collected the eggs one of our Barred Rock hens laid. I placed them in the incubator and turned them several times a day. At the end of the week, I slipped them one at a time under the Silkie's soft body, each time removing one of the older eggs. She accepted them easily.

We were sure the eggs she had been sitting on would be too far gone to eat or to cook and feed to our animals. Out of curiosity, we cracked the first egg. Inside was a tiny pink bird with an enormous eye. It had perfectly formed, miniature legs and a complicated network of blood vessels connecting it to its life support system.

I was horrified and fascinated—horrified because I felt as if we had just performed an abortion, fascinated because of the perfection of the tiny creature.

We should have immediately tucked the rest of them back under the Silkie, but we figured the first egg was an aberration and the rest would spill sulfurous putrefaction upon cracking.

The second egg held a little black bird, no more than a week away from hatching. The third held a speck of a chicken, mostly eye. And so it went, with all seven eggs. I was mortified. Had we left the eggs under the Silkie, the first chick would have pecked free of the shell long before the others were ready. Maybe the others would have died, as the Silkie turned her energies to the newborn.

More likely, they would have hatched in due time and thrived. The process was unfolding perfectly until we intervened.

Three weeks later, my husband went out to do the morning chores and found seven little chicks under the Silkie. The tiny hens had the colouring and stripes of chipmunks. The wee roosters were black with white spots on their heads. They were a cross between a Rhode Island Red rooster and a Barred Rock hen. The Silkie fiercely protected her little brood. Friends told us the chicks would die because they were born in mid-winter. Their mother ignored the warnings. When the little ones strayed too far, she spread her wings and called to her adventuresome brood. They scurried back to the shelter of her warm body and snuggled contentedly.

All six survived to adulthood. They followed their little white mother everywhere until they became too grown to heed her calls.

Sometimes our inexperience took a toll on our growing menagerie. Fortunately, the animals were generally healthy and patiently taught us what we needed to know to keep them that way.

Chapter Ten
BUT WE WILL HAVE TO KILL THEM

The old truck rattled down the driveway, sounding as if it were held together by wire and duct tape. A grizzled, grey-haired man climbed out of the cab. With minimal small talk, the poultry dealer asked to see our setup. We led him to the barn, where we had made a warming pen out of a child's swimming pool and hung a waterer and heat lamp above it. He gave simple instructions on caring for baby chickens and told us what we would need as they grew.

Then he headed out to the cheeping truck bed, piled high with boxes of chicks. He handed us one and walked back to the barn. We emptied the fuzzy creatures into their new home, paid the poultryman, and began serving chick feed to the little travellers.

The poultryman lingered for a bit. Satisfied our makeshift setup would keep the chicks alive long enough to shift blame to us if anything went wrong, he walked with us back to the truck. Before climbing in, he turned around and said, "You should buy some turkeys."

The chicks were already a major concession on my part. I had not even wanted a pup. I had been reluctant to start raising chickens for slaughter, yet here we were with a hundred tiny creatures counting on us to be quick studies in poultry care. Turkeys were out of the question.

"Payback isn't as quick as with the chickens, but profit's good," the poultryman insisted.

"We aren't set up for turkeys," I protested.

"Have some cardboard?"

Before the poultryman rattled out of the driveway, we had built a makeshift pen for a flock of turkey chicks. Maybe there were only fifty, but I remember them as another responsibility suddenly landing on our shoulders. Worse yet, in a few months, those turkeys and chickens would be big enough to truck off to the slaughterhouse. Unlike our auction-bought laying hens, these birds had no other purpose on the farm than to end up on people's plates.

The high cost of operating a small farm overcame my scruples. I pushed down my discomfort with raising animals intended for dinner. I began mentally calculating profits as the chicks peeped, pecked, and pooped.

Then the first baby chicken died, probably smothered on a cold day as the chicks crowded under the heat lamp—or maybe trampled while gathering around the feeding trays. Whatever the reason, it was a flattened, fuzzy cadaver. In my naiveté, I had assumed every chick would survive to adulthood. I was devastated.

Even after improving their quarters, more little ones died. I have no memory of how many succumbed to whatever kills chicks. Most thrived, but each loss forced me to recalculate the difference between what we spent on chicks, feed, and equipment, and how much would remain after paying for slaughter and distributing them to customers.

We had an epic disaster when Liz and Clare Weir were visiting from Northern Ireland. The automatic waterer went on strike, flooding the chick enclosure. When we found them, the stronger chicks were clambering atop their drowning cousins, trying to stay above the flood. Many were already dead. Those still alive were wet and shivering.

We grabbed hair dryers and began blow-drying the fuzzy babies, desperately trying to save as many as possible. I vaguely remember one of us racing to the thrift store to buy more hair dryers. The majority survived, but the deaths were a heavy blow.

Chapter Eleven
FREE RANGE GUILT

The demand for free-range chickens was high. We had a waiting list as soon as people learned they could expect a supply. We just had to figure out how to raise the best possible chickens. We had heard about chicken tractors, those hen houses on wheels. They are pulled around a pasture, so the chickens are always on fresh grass. That gives them a steady supply of bugs and tasty weeds while they fertilize the soil.

That sounded ideal. We returned to our favourite Saturday morning entertainment, the auction. We were hunting for something we could turn into a chicken tractor. A small, fibreglass greenhouse was just the ticket. Richard bought wheels and created a rolling chicken house.

The more I watched the breast-heavy white birds, the guiltier I felt. They were the same stock raised by factory farms, where they spent eight weeks in crowded pens, their legs barely able to hold up that meat. Their baby beaks were snipped to prevent pecking. Their weak hearts often gave out. The usual schedule of hatching to hatchet was eight weeks. We were determined to give our chickens a better life.

We slowed the turnaround to twelve weeks. That gave them time to grow more or less normally, though they still looked wobbly. Every day or two, we rolled the coop to a fresh spot. The birds ran to us and clucked happily whenever we moved the chicken tractor or refilled the grain trays and waterers. They grew plump on their diet of grass, bugs, and grain.

From the first batch, demand exceeded supply. Scruples be hanged. They would top up the savings depleted by the cost of raising them.

Most of them survived long enough to make the trip to the slaughterhouse—or "slaughter shed," if I were being literal. It was a small, family operation. We would pull up to the doors, grab the chickens by the legs, and hand them to the fellow who attached them, upside down, to the conveyor belt. A quick stun, a slit throat, and they were on their way.

Their hell ended quickly. To his credit, the guy who did the stunning and slitting was quick and efficient. For me, however, the scene was straight out of a horror novel. I had gone from eating very little meat to raising it for slaughter. From knowing too little about animals to properly train a pup, to becoming a quick study in poultry husbandry.

The small slaughterhouse was clean. We had no reason to worry about the safety of what we were selling. Still, I was unprepared for the sounds, smells, and blood of a carnivorous business. I felt like a quasi-vegetarian gone rogue, a killer on the loose.

We parked the old Ford truck in the chicken pasture, lowered the ramp, and scattered feed on it. The trusting birds pecked their way up the ramp and rooted around on a bed of clean straw, searching for tasty bits. We lifted the ramp and drove off to the site of their demise.

One day, a customer asked to see the slaughter process. We shuffled, hemmed and hawed, and found an excuse to keep her from witnessing the ritual of sacrifice. Most carnivores prefer only knowing what they see at the butcher shop or supermarket. Plastic-wrapped meat does not bleed or cry out in fear. Live animals do, however clean, quick, and careful the process that ends their lives.

I was never entirely peaceful with it, a sign of my status as a reluctant farmer. But I enjoyed their company while they grew and fattened. I blessed them for the profits they delivered. When we served one of them for dinner, I blessed them again.

At Auchinachie Farm, they were part of a cycle. If we put them on a previously grazed by sheep, they would scratch and distribute the manure left by the woolies and happily munch the grubs that burrowed into the soil. A sheep-cropped field turned lush and green in their wake.

The "meat birds" would never have been candidates for Mensa, but they were cheerful field workers. It was the first sign this farming venture might actually yield a small profit.

Chapter Twelve
MAMA CAN'T OUTWIT CROWS

After our initial experiences with poultry and sheep gave us a degree of confidence, we decided to focus on heritage breeds. That way, our small farm could at least contribute to the gene pool of endangered livestock.

Of course, that was only a broad guideline for our decisions. Take the black and white Muscovy ducks, for instance. Native to Mexico, Central, and South America, the red-wattled birds were a popular breed among the small farmers of Vancouver Island. They were not an endangered species, but they were good workers on the farm. We bought a few at auction and watched garden-munching slugs become duck feed.

They were confident creatures who wandered at will. They could have flown or waddled anywhere they wanted but settled happily into a small part of what was beginning to resemble Old MacDonald's Farm.

When one of the females went broody, she dug a nesting place with her bill. She plucked feathers and down from her breast to provide a warm, soft home for the eggs she was laying. She left them only for food and water or to relieve herself.

I was in my mid-40s and had never come closer to a hatching bird than the occasional documentary. I was enchanted. The duck was tolerant. As the hatching day approached, I visited repeatedly throughout the day. On occasion, I could not resist reaching a hand beneath her feathers. I doubt she was thrilled, but she was patient.

Then came the tiny cheeps. Sanguine Mama became Fierce Protector. When we approached too closely, she would gather her chicks beneath her wings, hiding their yellow-and-dark-brown bodies. She never pecked us.

Like small but mobile offspring of any species, they never stayed hidden long. Each time she stood up to gather them, we caught glimpses of tiny beaks tapping cracks in the remaining eggs, gradually widening them until the chicks popped free into her sheltering wings. I was mesmerized.

We kept the little flock enclosed for a few days, hoping to keep them safe for at least a short while. Then we let them out. The farm world was full of wonders—and hazards. Mama Duck could not curtail her little ones' eagerness to explore.

In astonishingly short order, I learned that a tiny duckling is just the right size for a crow's meal. Letting them run free was like ringing a dinner bell. Mama Duck would waddle along, teaching her offspring to forage, but the line of little ones grew shorter and shorter. My admiration for those smart, cocky crows underwent a serious trial. No wonder a flock of them was called a "murder."

On the other hand, the ducks helped me understand the origins of three more idioms and reinforced a fourth. Those fuzzy little ducklings were a literal illustration of the first idiom, "getting one's ducks in a row." Mama Duck did her best to keep them in a line, but the crows were only one of her parenting challenges. Sometimes her fuzzy children wandered off, lured by a sight or a scent. Frantic calls went up from under bushes or out of swaths of tall grass. Mama Duck scurried in search of the strays.

As for the second idiom, water was a hit in baby-duck land. They dropped into the pond "like a duck to water" and paddled happily away. Rain was no bother, even at the fuzzy stage, leading us to understand the provenance of the third idiom. As they began to feather out, whatever rain that fell on them was mere "water off a duck's back".

The chickens had already taught me the beauty of "taking someone under your wing." Mama Duck reinforced the power of that expression. If the ducklings were cold, afraid, or in need of a nap, she spread her wings and gathered the little survivors under her warm, protective body. She gave them a small bubble of temporary safety in an uncertain world.

Chapter Thirteen
THE FARMER & THE EGG THIEF

Our suburban farm stretched over eleven acres, tucked between two schools. The passing munchkins were fascinated by our animals. They would pause at the fence with treats, hoping to pet a lamb or scratch a goat. One of our goats had backward-curving horns. He could stick his head through the paige wire fence to accept a proffered carrot, but he could never pull it back. He knew the drill and would wait patiently while a child ran ran breathlessly to the house, announcing, "The goat is stuck again!"

Phantom, our sweet but simple Border Collie, loved socializing with the children. They knew it and would call him as they walked to and from school. Several times a week, he would set off in search of them. We would have to walk down the street, checking the playground, laughing at the sight of him happily escorting a small human back home. The children quickly realized he could be used to their advantage—walking Phantom home became a surefire plan for a classroom timeout.

Occasionally, other animals succumbed to the lure of succulent greens on the other side of the fence. Children would run to the door to tell us a turkey had gone walkabout or a goat was munching a neighbour's flowers. We would drop whatever we were doing and chase after the errant livestock.

Then one of our Muscovy ducks laid a clutch of eggs in a space she hollowed out beneath the cedar hedge that bordered the road. Plucking down from her breast, she made a soft carpet for her future ducklings. We checked on the nest daily, watching for the first signs of hatching.

One day, we noticed an egg was missing. The next day, another disappeared. We suspected some egg-sucking creature was dining on duck eggs until the day Richard walked by the nest just as school let out.

As he stopped to count the eggs, he saw a small hand reach beneath the cedar hedge and wrap around a warm egg. Instinctively, he grabbed the hand. The hand's owner snatched it back, bolted down the road, and screamed as if pursued by a murderer.

After that, no more eggs disappeared. We never heard a word from the child's parents, but the thief had been caught in the act. Mystery solved.

Chapter Fourteen
HEROISM IN A SMALL, FIERCE PACKAGE

The chicken had little more going for her than her beautiful feathers, but that was enough. She was an auction chicken, a bantam hen with feathered bands of orange, white, and black. The little hen belonged to the popular Mille Fleur variety of the Belgian d'Uccle breed. Cheap and colourful, she would add a spice note to our mixed flock of laying hens.

We named her Millie[1], a name more handy than creative. Her mothering instinct was so strong that she became the foster mother for any orphaned or injured chick. She was already raising three Sussex-cross chicks when we introduced her to Turkey Baby.

Turkey Baby was the runt of a batch of turkey chicks. He arrived with an injured wing. As even newly hatched turkeys do when they see red, or anything else that rouses their curiosity, his mates kept pecking at the blood.

We rescued the little one, washed off the blood, and placed him in front of Millie. She glared at him and gathered her chicks. The poor little turkey was shivering from his bath and weary from his injury and the pecking. He walked the few steps to Millie, lowered his head, and pushed it against her breast.

We held our breaths, watching to see if she would accept him, banish him as an intruder, or attack. She had scalped an orphaned Muscovy duck chick we had hoped she would foster, so our hopes were cautious.

Millie eyed the baby turkey. Keeping her chicks under her wings, she inched forward, puffed her breast, reached out, and drew the little one into her family.

We had planned to return the turkey chick to his flock once he grew strong enough to fend for himself. Millie would have none of it. Turkey Baby was hers. No matter how large he grew, she was always his mother.

Early one morning, Millie screeched. I rushed outside to find her facing down a hawk. Though smaller than her fosterlings, she had sent them into hiding and stood alone in the open, ready to fight off the predator.

Another day, I looked out from my office to see one of our roosters trying to jump one of her chicks. Already three times Millie's size, the young turkey would have seemed untouchable—but Millie raced forward, puffed up her chest, and placed herself between her chick and the intruder. The rooster turned tail and ran.

One evening, I was raking rye seed into the bare soil destined to become a garden. Millie's little flock accompanied me, pecking at the seeds. As dusk approached, they began heading home. I looked around for Millie. When I didn't see her, I walked to the pen to see if she had retired early. There was no sign of her. I walked the route the family always took—through the herb garden, along the fence by the ram pasture, under the apple tree, across the front yard, along the road, and back to the maple tree.

In the gathering dark, I found all that remained of Millie—tufts of scattered feathers. Her grown fosterlings carried on as though nothing had happened. But I mourned her fierceness, her courage, and her gentle, unshakeable acceptance of orphaned chicks.

1. The full story is in a three-part, picture-book series: *Millie's Feathered Foster Family, Turkey Baby and the Hungry Hawk,* and *Turkey Baby Finds Her Magic.*

Chapter Fifteen
TRUST FROM A WOUNDED CHICKEN

Humans trip and break legs. We fall off bikes and dislocate shoulders. We skateboard down hills and crack heads. I wanted things to be different for our animals. They should never fall ill or be injured.

The responsibility I felt for them was new to me. I had been pet-free for decades and had never had children. Now we had a small farm populated by dozens of creatures whose health depended on our understanding and looking after their needs. We had a lot to learn about livestock management. In fact, the small-farm enterprise required a basic knowledge of mechanics, biology, botany, construction, engineering, soil management, business, and marketing. My respect for those who feed us soared.

One of my early lessons in inattention came from a chicken. She was one of the hens who had flown the coop, enjoying free run of our suburban farm. She could have wandered into any of the neighbouring fields, but she chose to stay on ours. What she did not choose was to remain in the hen yard. Each night she roosted in a tree. Where she hid her eggs, or whether she even laid any, remained a mystery. She was independent, an adventurer who clucked confidently and chirruped contentedly on her rounds.

One day, I saw her limping. I picked her up to investigate and found that one of her claws was caught in a string. The string had cut deeply into her skin. The bottom of the claw was filled with pus. My squeamishness kicked in.

She was a chicken who loved snuggles, so she did not struggle when I picked her up. Fortunately, Richard was nearby. As I held the hen against my body, he began carefully cutting away the string, revealing deep gashes where the swollen claw had encased it. Blood poured down her upraised leg.

Despite the pain we were causing, she rested her head against me and patiently accepted our ministrations. Only once did she wince and try to pull away.

When the string had been removed, her claw disinfected and bandaged, she needed a place to recuperate. We put fresh straw into a cage and tucked her in, safe from pecking by other curious chickens. She nestled down, exhausted from her ordeal.

Every few hours, I stopped by to bring her fresh greens, check her water and feed, and see how she was faring. While she recuperated, she was content to stay in her temporary sanctuary.

The wound healed quickly, with no lasting damage. The day we let her out, she resumed her wandering and tree roosting as if nothing more than a brief vacation had interrupted her rounds.

My dismay at overlooking her distress until the string had caused infection. She, however, carried no resentment. Once recovered, she was far more interested in hunting for bugs than holding onto resentment or fear.

Chapter Sixteen
CHICK ON THE LOOSE

"Do you have any chicks?"

The voice was unfamiliar, but when I learned the caller had found a turkey chick wandering in her garden, I said I would be right over.

I walked to the neighbour's yard and followed the sound of soft, unhappy cheeping. A turkey chick peered at me from the back of the chicken coop.

Days earlier, the neighbour's friends had spotted a small chick in her yard. They had caught it and put it in a box. The little escapee was frantic and had wings strong enough to carry it out of the cardboard prison.

Another chase and capture ensued. This time, the pursuers locked the chick in the coop and began canvassing the neighbourhood for turkey farmers. While they scoured the neighbourhood, the lonely little turkey fed and slept with the chickens but mostly cried dejectedly for its mates.

The wanderer was not quite three weeks old and barely six inches tall. It had been living in the barn where we housed our turkey chicks until they were big enough to range without fear of predators. From the barn to the neighbour's backyard was a long and treacherous journey for such a small bird.

We scratched our heads, wondering how a small chick had made such a big journey. Perhaps it escaped out the door while we were feeding other animals. From there it would have crossed the barnyard, pecked its way through garden, scrambled along the easement, and navigated a neighbour's horse pasture and woods. Then it would have climbed a steep embankment to the yard.

Maybe it had walked up the long driveway and out the gate. The route from there would have been down the road and past our rental cottage. It would have wandered across the easement, into the neighbour's driveway, past their chicken coop, beyond the excitable dog, and into the backyard.

There were countless variations of the possible route. All of them were long and dangerous for a turkey chick only a few weeks old. We could never know how frightening the journey had been. All we ever knew was that the little turkey was traumatized and wanted to go home.

The neighbour lifted the chick from the coop and placed it in my hands. It settled quietly, cheeping softly, as if somehow aware that the long, lonely adventure was over.

That was the last time the little turkey wandered. Other animals would set out on adventures in the future, but this tiny escapee had learned its lesson—and we had learned a little more about the resourcefulness and determination of our birds.

Chapter Seventeen

TURKEY LIBERATION FRONT

The poultry guy advised us to buy enough turkey chicks for Thanksgiving (mid-October in Canada) and Christmas. We did some quick mental math: If the birds were the right size for Thanksgiving dinner, they would be monsters by Christmas. So we decided to slaughter them all and bought a used freezer to hold any unsold birds until the December holidays.

Our first batch of turkeys was as overbred as the meat-bird chickens. Their yellow fuzz soon gave way to white feathers. Their breasts and legs ballooned to shapes I had never encountered in pictures of the native turkeys introduced to settlers by the first peoples on this continent. These birds redefined plump.

Once they outgrew their quarters in the barn, we planned to move them to a side yard, then out into one of the fenced fields. They would be a splendid sea of white. Visions of paying a few bills —the chicks, their feed, and other expenses—helped calm my usual anxiety about the economics of farming. The prospect of a bit of profit was energising.

They were sociable birds. Every morning, I sang to them as I checked their feed and water and added fresh hay to their litter.

The gentle birds gathered around, eyeing me curiously and chirruping softly. They walked in a slow circle, observing me from every angle. People had told us turkeys were ill-tempered. This flock lived up to none of the warnings. As chicks, they climbed all over my lap. Even as they grew, some still tried to snuggle on my knees when I bent down to check water, pour feed, or clean the pen.

Anyone who has raised turkeys has likely heard the myth that they will drown in a rainstorm. We suspected that might be true. The first test came when they were nearly full-grown. Fearing for their safety, we ran to their yard. Some were tilting their heads, observing crazy the wet stuff. None of them was drowning. Still, we frantically scooped each one up, put them inside the barn, and closed the hatch.

One afternoon I discovered a gang of half-grown birds clustered around one poor tom. They were pecking furiously at his bloody tail. I went ballistic. Standing in the middle of the flock, I shouted, called them cannibals, and threatened to wring their necks. I was grateful no one was there to witness my tantrum.

The victim stood in dumb silence, hardly moving, unable—or unwilling—to defend himself. I tried to herd him into an enclosure where we used for recovering animals. He wanted none of it. He began breathing heavily, running and flapping to get away from me. He was stressing the rest of the flock.

It took me a full quarter of an hour to finally get him to safety. I brought him feed and water and built a barricade along the fence to keep the others from pecking him. He wanted out. Better the pecking than the isolation. But I knew they would peck him until he bled to death, so I made sure he was safe and then left, upset with the scene, the distress, and my reaction.

Over the next two days, the turkeys gathered by the fence, in solidarity with their imprisoned comrade. He calmed enough to eat a little and allow his tail to heal, but there was no peace in the turkey barn. Being on his own upset him. His misery unsettled the other turkeys.

Several days later, I found the barricades knocked down. Somehow, the flock had opened the gate and freed the tom. None of them had been caught under a falling barricade. The injured tom had recovered enough to square off with another male, engaging in the posturing that would lead to breeding rights if they weren't all breast-heavy beasts destined for the dinner table.

I was awestruck and humbled. The overbreeding of those turkeys had surely dulled their native intelligence. Yet they had freed one of their own, finding no peace until they did. Except for the posturing of the two toms, the flock was quieter than it had been in days.

Later, a woman who had farmed in Alberta explained that turkeys peck at anything red. She figured our tom had probably injured himself, and the others were pecking at the interesting colour. Since we had never seen cannibalism before, she was likely right. Regardless, the "Turkey Liberation Front" taught me new respect for the birds.

They were social creatures, with the usual share of squabbles and alliances. Unfamiliar objects and bright colours caught their attention. Sudden noises set off a wave of gobbling. Like most animals outside the human sphere, they did not hang onto slights or grudges. Any spat was short-lived, any hurt quickly forgotten.

Though bred for an excess of white meat rather than brains, they were still fun to hang out with. We learned out how to load them into the truck without scaring them and took them off to become the centerpieces for Thanksgiving dinners.

Over half of the birds had been spoken for by the holiday. We laid a grain trail from pasture to truck, and the turkeys cheerfully climbed the ramp, oblivious to their impending fate. Our customers were thrilled. The birds had been wandering the fields for weeks. Their meat was firm and flavourful. At 5-8 kilos dressed weight (feathers and inedible bits removed), they were perfect for the upcoming Thanksgiving gatherings.

The day they were ready, a steady stream of customers dropped by the farm to retrieve them. They were thrilled. So were we. Only one customer failed to turn up. When I called her, she said she would pick it up the day before Thanksgiving. I thought she would be dismayed when I told her we would have to freeze it. Her reply changed my relationship with frozen meat from that day forward. She planned to pop it in the oven in its icy state. Training as a dietitian had taught her how to cook frozen meat safely. As the bird cooked, the moisture usually lost in thawing was reabsorbed.

We stored the birds intended for Christmas in the freezer and dreamed of what we would do with the profits. The smell alerted us to the demise of that daydream. The freezer's failure made us reconsider the wisdom of dispatching the birds while they were still oven size. Next year's Christmas dinners would wander the fields until their final journey, no matter how large they grew. (They ballooned to 10-13 kilos, but we found a ready market for them.)

The farm seemed entirely too quiet without their gobbling. The following year, we raised bronze turkeys, instead of buying chicks of the overbred whites. That is when we learned the next lesson of turkey raising. Bronze turkeys can fly.

Chapter Eighteen
SO THAT'S WHY THEY CLIP THEIR WINGS

By the year after the freezer fiasco, we had adopted a plan: raise only "heritage breeds and seeds." Our eleven acres were too small to qualify as a serious farm in most circles, but at least we could contribute to preserving genetic diversity. Watching the white turkeys clumsily wander the barnyard had felt like participating in some mad scientist's nefarious scheme. So we ordered bronze turkey chicks. Closer to their wild ancestors, they have smaller breasts and bigger ambitions, which we soon discovered.

The fuzzy little peepers who arrived in a cardboard box had never known any mothering. We tried to fill in, and they quickly bonded with us. Whenever we showed up to feed or water them, they gathered around, forgiving us for our inconstancy. When I sang to them, they gobbled quietly. Between our visits, they had each other, and there was no repeat of the bloodthirsty pecking that had marked our first turkey-raising experience.

As they grew to knee height, we learned to stop wearing rings, watches, jeans with brads, or anything shiny. Followed by a flock of friendly, curious turkeys, pecking at bright bits on our clothing, I finally understood how the white birds might have ganged up on the injured tom. Their intent would not have been malicious, despite the brutal results.

As they grew, their bronze feathers shone in the sun. We pastured them, confident the grass, bugs, and fresh air would make them healthier and tastier. That is when we discovered their free-range ambitions.

Without the heavy breasts of their white cousins, the bronze birds could fly. Like their wild relatives, they had their eyes on the far horizon. They took to the air as soon as their wings were strong enough. We had six-foot fences, at least twice as tall as the white turkeys could have escaped. Those fences were mere midway perches for the bronze turkeys. Before long, they were flying wherever they pleased. By day, they grazed peacefully in the pastures; by night, their wild genes sent them to the trees.

Ours was a suburban acreage, with the ever-present possibility of rogue dogs attacking the turkeys. We brought them into the barn and the attached yard to keep them safe at night. Wrestling them out of trees was not out of the question for a couple of middle-aged farmers. Clipping their wings was not something we discussed. Neither was herding them home. They had to return on their own.

We quickly learned one of the primary rules of living with livestock. A bucket full of yummy grain is an irresistible siren's call. We have no videos of those treks from pasture to barnyard, but I can picture them perfectly: a big white bucket bobbing toward the barn, followed by a flock of turkeys gobbling excitedly, those closest to it pecking at the treat-filled container.

Ah, the bucolic joys of rural living—just us and a few dozen turkeys, noisily heading home for the night.

As for the anticipated profits, we spent some of them doubling the height of our fencing. At least it kept the turkeys safe.

Chapter Nineteen
BROWN THUMB GARDENER

Until we moved to the farm in June 1991, my experience with gardening had been limited to the narrow confines of urban yards. Suddenly we had space for a large flower garden, a larger herb garden, and an even larger vegetable patch. My husband was working full time at the newspaper. He used our small tractor to prepare the ground, but the task of creating the gardens fell to me.

I was clueless. The Internet was years away, but the former librarian in me started digging, physically and metaphorically. I had been more of a brown-thumb gardener than a soil tiller. Things died under my care. Vancouver Island's mild climate was the gardener's friend. For me, it was a taskmaster, requiring year-round planting, weeding, watering, and harvesting.

By working every free hour, I created a colourful display of annuals and perennials to greet visitors who arrived at the farm. I ordered open-pollinated seeds and started the plants in our small greenhouse. I learned how to save and store seeds. Bees buzzing in our gardens were my friends, working alongside me. The vegetable garden provided a bounty for our table and plenty more for canning and freezing. The whole cycle seemed miraculous.

Meanwhile, my storytelling work was still in the States. In August of our first year on the farm, I drove to Seattle to stay with a friend and commuted to Pacific Lutheran University, just outside Tacoma. The students in my storytelling class, all women teachers, were the epitome of enthusiasm, embodying the original meaning of that word, divine inspiration.

Back on the farm, I discovered the ten-day absence had been enough for the weeds to stage a takeover, out-competing my carefully planted vegetables and flowers. The first wheelbarrow I piled high with weeds almost seemed like a blessing. Our laying hens were thrilled with the fresh greens, and I knew the weeds would give their eggs hard shells and bright yellow yolks.

I emptied the wheelbarrow and returned to the vegetable patch. A light rain fell and fell, soaking me through in an hour. I was muddy, discouraged, and philosophically aware that the plants I was discarding were hardy natives—but my shoulders and heart ached for the domesticated ones I had nurtured.

Looking up from a patch of lamb's quarters and chickweed, I saw only weeds. They choked the pole beans, shaded the tomatoes, wound around the peas, and smothered the herbs. I lost heart and wished I were back in the city. I castigated myself for being such a poor gardener.

The rush of wings pulled my eyes skyward. A flock of ducks flew past. They were our white, blue, and green Muscovies, returning to the pond. My gaze moved from them to the double dome of Mt. Prevost. *Swuqus*, as the Cowichan people called it, the place where life began. From the stony swelling of the domes, my eyes moved down to the wooded hills, then dropped to the fertile fields around me.

Concentrating on the weeds, I had forgotten the beauty that surrounded them. I returned to my task with a different eye and a lighter heart. As I pulled chickweed from around the pole beans, I set some aside for salads and found enough ripe beans for dinner. Hidden among the pea plants was a handful of sweet pods. Three tomatoes, their ripening hastened by a start in the greenhouse, hung red and firm. When I pulled aside borage leaves to weed beneath them, I discovered a forgotten chervil plant, its seeds ready for late-summer sowing.

I had been overwhelmed by the intruders, yet even weeds held promise. Those vigorous plants thrived where I wanted others to grow, and moving them to the chicken yard allowed them to be appreciated. They became a metaphor for the weeds in my inner life—memories and regrets that threatened to choke growth. Offering them to my inner "chickens," I could feed them to wisdom, laughter, and perspective.

The image made me laugh and bless Jeanne, my Spotted Chicken friend. With a happier heart, I returned to weeding. This time, I thanked the plants I was pulling out. They had their own beauty, and so did I.

Like most beatific realizations, mine was temporary. The weeds kept growing. I kept being a reluctant farmer. But the good times generally outweighed the uncomfortable lessons.

Chapter Twenty
INTERFERING WITH TRUE LOVE

We had heard geese mate for life. We had no idea how deeply they bonded until we learned the hard way or, rather, until they learned the hard way how callous humans can be.

We were looking for geese to add to our Old MacDonald's Farm menagerie. I was hesitant at first. My only prior experience with the species was when a neighbour's backyard goose attacked my three-year-old self with her powerful wings.

Still, geese can be excellent watch animals, setting up a ruckus if intruders come around. The grass they eagerly graze grows lush in their wake. Besides, what is a mixed farm without geese?

So off we went to the Saturday auction and acquired three. Two were Toulouse; the third was African. One of each breed turned out to be male, leaving one Toulouse female and an upcoming spat.

For a while, they wandered as a trio of pals. Then the female felt the urge to lay eggs. The two males squared off for mating rights. We placed our hopes on the Toulouse, but the more aggressive African won the face-off. The female and the winning male became inseparable, while the Toulouse male remained an outsider.

We had built them a snug hay-bale house, which they

completely ignored until after the battle. The next day, the Toulouse female began building a nest inside it. When she was happy with its size and comfort, she laid an egg and tucked hay around it. Over the next three days, she laid two more. Then she nestled down on them and prepared for motherhood.

The African stood guard, hissing when we came near, rushing angrily after the Toulouse when he dared approach. The losing male would not be deterred. Whenever the African wandered off to graze, the Toulouse would hurry to stand guard until he returned.

We knew this first batch of goslings would be mixed breed, but we were determined to have Toulouse goslings next time. So we rounded up the African and sold him at the auction.

The female was bereft. When she was not sitting on the eggs, she wandered disconsolately, searching for her missing mate. The Toulouse male hovered nearby, ensuring nothing bothered the eggs until she returned.

He got no thanks. They had been pals before the mating season. Now they were strangers. She tolerated him, but he was no substitute for the African. Perhaps the trio had already sorted out their relationship before the hissing, flapping fight settled it in our eyes. Whatever the case, we had disrupted a real relationship, and the female's grief was visible.

In time, three goslings hatched. The Toulouse male followed them like a worried uncle, protecting them as if he had fathered the youngsters.

Eventually, the two became a pair, but the African exacted his revenge. Although the goose repeatedly laid eggs during the remaining years she lived with us, she was never again a mother. The Toulouse male tried his best, but he could not do his part. The female would sit for weeks on sterile eggs before abandoning them. We would remove them and add them to the compost pile. Then her cycle would begin once again, always with the same outcome.

The geese, the chickens, the piglets, and the sheep all made me yearn to share the farm and my uncertainties about farming with my mother. Although dementia was gradually loosening her hold on life, I still dreamed of bringing her to the farm. I had grown up hearing stories of her happy childhood on a small Nebraska acreage. I thought spending time on ours might comfort her, perhaps even revive some of her memories.

It was not to be.

Chapter Twenty-One
THE ORPHAN'S CRY

At some point in our lives, we all become orphans. Unless we die before our parents, we live a chunk of our years as motherless children. Fatherless too, but mine played only a small role in my early childhood. I was in my 40s when he died. I never really knew him.

Mother was my best friend. I could tell her anything without risking a dent in her unconditional love. That continued after I graduated from high school and through my university years. It remained strong when I married and moved far away.

She moved from Idaho to California to be near the grandchildren. I moved to Seattle and New York and spent shorter periods in Oklahoma, Texas, Germany, and the Netherlands. We still had a psychic connection. When the phone rang, we knew if the other person was calling. When dementia began stealing her away, I wrote to a friend:

"My mother is missing and may never come back. Oh, she is there somewhere. Somewhere in the soft interior, protected by thick walls of memory. She sits there waiting, endlessly waiting. For death? For the return of people whose love she misses—parents, siblings, old friends?

"Mother no longer knows her address. She no longer lives there anyway. She has checked out. Moved on; moved in, so far in that outside voices reach her only dimly.

"She still functions. Feeds the cat. Bathes when Eric and Carla remind her. Changes her clothes when they suggest it. She never cleans her apartment or cooks a meal, but she goes daily to the senior center. She potters about, offers help, finds comfort in the known, in the company.

"And I wrestle with sadness and confusion and guilt at not being closer. All of the burden of looking after her has fallen on Eric's shoulders. I keep wishing we could somehow bring her up to the farm. She always loved digging in the garden. Maybe she would find solace in it again. Or maybe it would just frighten her if she couldn't figure out the difference between a weed and a cultivated plant."

Two months after I wrote that letter, an aneurysm brought on sudden death. She had remained independent far longer than most people with dementia. A kindly landlord checked on her regularly. So did my brother and sister-in-law. She lived in a unit of a horseshoe-shaped apartment complex. The elderly women in the one-bedroom units knew each other's habits and sounded the alarm if anyone's blinds were not up by a certain time or a newspaper was not picked up.

Despite all that care, life for Mother became too confusing for her to be safe on her own. Eric bought her a microwave with two settings, on and off, thinking that, at least, she would be able to heat her meals. Even that proved impossible. She would stare at the machine uncomprehendingly, not understanding how to open the door or turn it on.

My brother and sister-in-law found a care home for her. She was to move at the end of June 1992. On the seventh of that month, a lung aneurysm ended her life. I believe she had the internal strength to choose to leave rather than accept incarceration.

I was devastated. We never outgrow the need for a wise, loving elder, but there comes a moment when they must leave us. I was just past my mid-40s, still finding my footing as a second wife, a new immigrant, and a reluctant farmer. More than ever, I longed for my mother—not the fragments left behind by Alzheimer's but the strong, gentle presence who had always been my anchor.

I flew to California to clean out her small apartment and attend her memorial service. The apartment made me laugh and cry. Everything spoke of a mind no longer making connections along customary paths. I found utensils in the freezer compartment of her refrigerator, old papers in the bathtub, and soap in her purse. I found the receipts she had kept for my father's alimony payments. There were only two, one for $10, another for $25, both sent shortly after their divorce.

At the funeral, people told stories of how she had touched their lives. Some talked about her many mystical experiences. I remember one of their stories, of Mother's witnessing a car hurtling toward another vehicle. She saw the hand of God reach down and stop the car. The other driver had no idea how close he had come to being killed.

When we walked out of the hall that night, the full moon was just being hidden by an eclipse. Mother had been my full moon. Her eclipse was permanent.

When I returned to the farm, Richard showed me the rose bush he had bought in her honour. He remembered that Mother often said if she had one last dollar, she would buy roses instead of bread.

As I looked at the rose bush in the heart of the future flower garden, I heard a few lines of a hymn running through my head. I had been among the unchurched so long that hymns no longer came unbidden to my mind. I weeded, dug, watered, and planted, ignoring the melody until I realized the words were about courage, about knowing we were not alone.

When I awoke the next morning, another old hymn was going through my head, "God will take care of you. Be not afraid." I wondered then. I wonder now. Was that you, Mother, sending comfort, wrapping your arms around me, reassuring me you were still there, still loving me, still listening?

Thank you for being my closest friend. You were the best mother I could ever have hoped for.

Chapter Twenty-Two
DOESN'T EVERYONE HAVE A PET BAT?

My office was in an upstairs bedroom that looked out over the fields to the distinctive shape of Mount Prevost. An old, double-hung window could be pushed down from the top, bringing cool air into the room without chilling my typing fingers.

Early mornings were my most productive time, before feeding livestock, weeding the garden, or watering the pigs. One morning a bat flew in through the open window. He navigated over the desk, around the bookshelves, and into the closet where I stored office supplies. There he stayed.

Curious about what he was doing in the closet, I tiptoed to the door. The bat hung quietly, wings tightly folded, hanging from a shelf in the far corner.

After breakfast and chores, I returned to the computer. Before settling in, I checked the closet. The bat was still hanging there.

In the evening, as the sun set, the bat emerged from his closet cave and flew out the window. I checked the closet and was surprised to find no bat droppings.

The next morning, the bat returned. This time, he flew directly over my head and right into the closet.

I was surprised by the solitary nature of this little visitor. Maybe he didn't get along with his bat brothers. Whatever the case, I found odd comfort in his daily return, especially since he never defecated in my closet. I've been told that it is not possible, that bats always pile up the guano wherever they roost. This one did not, or he would have quickly made himself *Myotis non grata*.

A month passed before my husband discovered my little companion. I can't remember if I finally confessed I had a visitor or if he discovered my friend while looking for office supplies. I do remember the warnings of fleas, rabies, and soiled supplies that ensued. So did a closed window.

I had known that this little guy's sojourn in my closet could not last, but I suffered guilt pangs until he stopped trying to claw his way into the window each morning. I was enjoying the little creature's quiet presence and wondered if he would be all right.

Worrying was something I had honed to a fine art. Richard used to tease me with the phrase, "75/25." He meant that 75% of the things I worried about never happened. He was right. Remembering the bat visitor now, I ponder how many of my fears during the farming years were like the scary stereotypes about this small, winged insect eater. My imagination turned them into frightening, blood-sucking monsters. Many of them actually brought gifts, and I do not mean guano.

Chapter Twenty-Three

FRED EAGLESMITH & FARM HEARTACHES

We had been living on Auchinachie Farm for a year when Richard decided to step down from editing the local newspaper. His commitment to honest reporting sometimes led him into sensitive territory, earning the respect of readers while occasionally ruffling feathers behind the scenes. Like most small-town papers, this one depended heavily on advertising revenue, often from the same organizations that appeared in its pages. It was a delicate balancing act between editorial independence and financial necessity. The time came when stepping away felt like the right choice.

I was still performing and leading storytelling workshops, but the steady market I'd built in Seattle was no longer within easy reach. My income had dropped, and I was learning to navigate a new landscape for my work. Ever the skilled wordsmith, Richard was well acquainted with the unpredictable rhythms of freelance life. Between our modest stipend as music directors for the Cowichan Folk Guild and the work we poured into the farm, we found ways to keep things going while adjusting to a simpler, more uncertain kind of life.

Small farms contribute far more to a region's quality of life than their size might suggest. They can be profitable, with enough acreage, a solid business plan, a reliable retail market, and a generous dose of luck. Weather, shifting consumer tastes, and the countless things that can happen to livestock all play their part. Success also depends on enthusiasm and commitment from everyone involved. In our case, one of us was a reluctant farmer.

The imbalance between income and expenses sent my worry meter into overdrive. In my mid-forties, I began to picture a future with little left in savings by the time retirement arrived. Still, we wanted to learn all we could and connect with others who shared our hopes. With that in mind, we helped organize the Cowichan Smallholders, a group of people working hard to strengthen both community and livelihood through whatever their small acreages could produce or offer.

One of our dreams was to turn Auchinachie Farm into a place for performances and workshops. That seemed within reach for two people with a long history in front of audiences. We scheduled our first concert even before we had a proper venue, enlisted the Cowichan Smallholders as co-sponsors, and hoped for the best. Luck brought the right performer our way.

A fellow smallholder had introduced us to the music of Fred Eaglesmith after hearing a song on CBC that stopped him in his tracks. The lyrics moved him so deeply that he had to pull over to the side of the road. The song was "Thirty Years of Farming," from Eaglesmith's *Indiana Road* album. It captured the heartbreak and endurance of anyone whose farming dreams hung by a thread. It told of a man who had worked his land for three decades, only to lose it to foreclosure. On the day of the auction, he and his wife stood by helplessly as everything they had built slipped away.

We were all new to farming, living with equal measures of hope and anxiety, and we felt certain others would relate to Eaglesmith's songs. We called his agent to see if he might be passing through.

He was, and in the days before the concert, friends from around the Cowichan Valley appeared at our farm with tool belts, hammers, and determination. We were still finishing the stairs to the barn loft when Fred arrived, along with Willie P. Bennett and Ralph Schipper.

With stage lights and a sound system in place, the old barn loft became a perfect stage for Fred's soulful lyrics and the trio's fine musicianship. Now and then, a sheep would call from below, as if joining in the chorus.

In that rustic loft, with the crowd's laughter, the music's energy, and the animals' answering cries, it felt as if everything might just work out. That night in 1992 was touched with magic.

Chapter Twenty-Four
A CORKER OF A PORKER

Even before they arrived at Auchinachie Farm, two pigs taught me to admire their escape artistry.

We bought our first pair of weaners—piglets just weaned from their mother—from a nearby farm. They had been raised in a large, airy barn where clean concrete was the only surface they knew. We came prepared with a sturdy carrying cage, which we placed by their pen, ready to usher them into their new ride.

That was the plan.

The moment we opened the gate, one piglet spotted an opportunity and shot past my leg. While Richard pointed out that I wasn't exactly a star pig catcher, the second one darted cleanly past him. Score two for the pigs.

Richard dashed to close the door at the end of the corridor, calling back to me to set up a barricade. We feared that if they reached the big world outside, they might add to the feral pig population on Vancouver Island. With the barn door secured, plywood for barricades in place, and the cage planted in the middle of the corridor, we gradually narrowed their escape routes until the only option left was the carrier.

By some miracle, it worked. We slammed the cage door shut, fastened it tightly, and hoisted the squirming, squealing pair into the back of the pickup. The sound of a scared pig could convince anyone that serious animal abuse was taking place. Fortunately, by the time we reached the farm, they had calmed enough to resume plotting their next adventure.

As enthusiastic novices, we wanted to document everything. Richard sent me into the house for the camera. When I returned, the piglets had broken out of the cage and were rooting through the straw on the pickup bed.

We parked by their new pen. Richard grabbed the hind legs, I the fore, and we lifted the first small pig into its new digs. It fought and screamed bloody murder the whole way. We repeated the process for the second, with the same reaction from the piglet.

The moment their sensitive noses touched earth for the first time, they were transformed. They tore into the thistle- and burdock-choked future garden with instant porcine bliss, flinging dirt and weeds in every direction.

They crisscrossed the pen, limbs flying, snouts digging, snorting in piggy pleasure. Occasionally, one would stop long enough to glance our way, as if reconsidering our place in their world.

The pigs were Richard's idea for preparing a space for gardening, part of a plan to clear and fertilize the garden plot. Digging by hand would have taken weeks, and a rototiller would only have spread the thistle and burdock further.

Watching two joyous piglets plunge their snouts into the earth was enough to dissolve my resistance to expanding our menagerie. In minutes, they dug up more of the garden than a friend and I had managed in an hour, and they did it with obvious pleasure. The roots of sturdy weeds lay scattered and easy to remove. Even that proved unnecessary, as they delighted in the taste of the succulent roots, the prickly nettle leaves, and even the sharp-edged thistles.

They were corkers, those porkers. Within days, they decided we were useful creatures. Not only were we providing them with succulent fare and comfortable quarters regularly refreshed with straw, we also turned out to be easily trainable. We had fingers adept at scratching behind ears or beneath chins and hands willing to give belly rubs.

I had never been around pigs and knew nothing about their sociable natures. They were the smartest and friendliest creatures on the farm. They were also good teachers.

They taught us to think differently about the pork chops, hams, and bacon lying in neat, plastic-wrapped packages in supermarket coolers. We learned about farrowing crates, where artificially inseminated sows spend the final days of their pregnancies and the month after giving birth, confined so their piglets can nurse safely. (Much later, we learned just how easily a piglet can be crushed by its mother.) We read about tail docking, castration, crowding, antibiotics, confined boars, and rivers of waste that plague industrial farms.

Our pigs, at least, had a good life—earth to dig, clean straw to bed down in, and daily doses of affection. When their time came, we took them to a small family-run butcher who understood the importance of calm handling. He kept the animals long enough for the stress of travel to fade before quickly and humanely dispatching them. They came back to us as hams, bacon, and tenderloins—some of the best we, or any of our customers, had ever tasted.

I never fully grew comfortable with raising animals for slaughter, but those pigs were my first real teachers in understanding the food system. Years later, that experience would guide me toward a new path—helping communities create gardens, kitchens, and food policies that contributed to their health and quality of life.

Chapter Twenty-Five
A PIG LOVER & GODDESS WORSHIPPERS

The pigs arrived shortly before our first bed-and-breakfast guests. Someone with porcine experience told us raising them was simple. We just needed a few strands of electrified fencing and a power source. Let the pigs touch their moist snouts to the wire a few times, and they'd learn to stay put.

The problem was that no one told the pigs.

When they felt the sting of the fence, they charged forward, snapping the wires and disappearing into a tangle of blackberries.

The piglets were faster and more agile than we were. They provided hours of unintended exercise before we finally persuaded them that the safest place was behind the electrified fence.

Within a day or two, they had settled into their new quarters. Once convinced that the humans bringing feed and water were reasonably trustworthy, they turned friendly and inquisitive.

Whenever we appeared, they stopped what they were doing to see if we had anything interesting to offer. They tossed onions and citrus to the side. Apples, melons, and leftover garden produce made them squeal with joy. What went in one end as kitchen scraps came out the other as a major contribution to garden fertility.

Mixed with a bit of the earth they loved to turn over, their manure became nutrient-rich compost for our gardens. Even when they were not hungry, the pigs were still happy to visit. In quick order, they trained me to scratch the exact spots that made them groan with pleasure.

Their vocabulary was more complex than I had expected. Our pink pals had a wide vocabulary of snuffles, grunts, snorts, squeals, and sighs. A frightened pig could split the air with a squeal, but a belly rub brought deep, contented groans. New people or new foods made them wiggle their snouts and snuffle with curiosity. They were endlessly entertaining, and I spent a lot of time getting to know their quirks.

Then our first bed-and-breakfast customers arrived, a couple from the States who had come to participate in a goddess-focused conference. "Gaia" rolled easily off their tongues, as did the instructions they gave us about their personal care and feeding.

We were new bed-and-breakfast hosts and had not thought to ask them about dietary restrictions. Although I had spent a long time as a quasi-vegetarian, I had no idea how to feed vegans. Eggs, butter, and cream disappeared from my planned menu. While our guests slept in, I dashed out to gather ingredients for a plant-based menu I hoped would impress.

Ten o'clock passed. Eleven. Noon.

Farm chores could not wait, so I kept running back and forth between the house and the fields to see if our visitors had surfaced. Around two o'clock, with the chores done, I paused to scratch the piglets' bellies. I was trying out some of my best piggy language when I heard an unmistakable exclamation of disgust behind me.

"Ewww. How can you stand to touch such dirty creatures"?

The goddess worshipper had finally descended from her beauty sleep and was staring at me with icy disdain.

I greeted her with as much warmth as I could muster and invited her to have breakfast in half an hour. Around two-thirty, her partner wandered downstairs, and I served them both. They picked at their plates with the clear message that ours would not be their pilgrimage spot of choice next year.

It surprised me that someone who spoke so reverently about the interconnected web of life could be so revolted by two of its most intelligent and endearing members. The pigs, for their part, continued to give me endless hours of amusement. Mostly, they stayed within the boundaries of whatever area we staked out for them, but one rule of raising livestock is that fences are just suggestions. When a determined animal decides the feed is better on the other side, the fence is a temporary impediment.

I was working at my desk, with its stunning view of Mt. Prevost and a full panorama of our little farm, when Richard came running up to the house. He motioned for me to open the window.

"The pigs are out," he shouted, breathless from the chase.

I did not help matters by taking my time. I ambled downstairs, walked to the barn, filled a bucket with feed, and headed toward the commotion.

"Hurry up. You go around that way. I'll go this way," Richard called, circling to cut them off.

But instead of joining the pursuit, I stood where the pigs could see me and shook the bucket.

Food! Oh, piggy joy. They bolted from the blackberry bushes and trotted straight toward me. The prospect of extra grain was seductive, so they followed me back to their pen, where they stayed put long enough for us to repair their fence.

The feed bucket proved its worth repeatedly. The promise of a treat is a powerful incentive for most farm animals and a time saver for the farmer trying to coax a runaway back home.

We still had much to learn. In the weeks that followed, our eagerness to expand the menagerie made us easy marks for a shady character who sold us a goat and a small black ewe.

Chapter Twenty-Six
A SWEET WAY TO GO BROKE

One thing we had in abundance was blackberries, sprawling over fences and laying claim to every open space. We had heard the stereotype: goats will eat anything. Goats and blackberries sounded like the ideal match.

We were ready to set out for the auction, determined to bring home a goat. A friend who knew our plan put us in touch with someone selling a goat and suggested we meet before the auction began. We congratulated ourselves on being shrewd. The seller's truckload of sheep and a goat was already bound for the sale barn. As newcomers, we were sure we would save money by buying directly from him.

We showed our cards too soon. The seller quickly learned why we wanted a goat and said he had just the one for us—a sweet-tempered Angora. Knowing we had an interest in heritage breeds, he mentioned he also had the perfect sheep: a little black Karakul, a rare, fat-tailed breed common in Africa and Asia but seldom seen in North America.

We fell in love with both of them, handed over our cash, and drove back to the farm with visions of cleared brambles and beautiful wool dancing in our heads.

Once home, we quarantined our new animals, just to make sure they weren't carrying anything that could spread to the flock. We named the goat Angie. She was enchanting, all silken curls and gentle eyes. We ran our fingers through her fleece and admired our new "land-clearing specialist." When we were confident she was healthy, we led her to the blackberry patch and watched her tuck in with enthusiasm.

Mission accomplished, or so we thought.

Minutes later, plaintive bleating rose from the brambles. Fearing a predator, we ran to find Angie in the middle of the patch, hopelessly ensnared. Her luxurious curls had caught on every thorn. She bleated pitifully while we untangled her. We gently led her back to the edge and went on with our work—until, ten minutes later, her cries rang out again.

Our perfect bramble-muncher was a perfect misfit. Her silky coat caught on the tiniest thorn. Any benefit she brought us was more than offset by the time it took to free her. We fired her from blackberry duty and promoted her to pet.

The Karakul's troubles were more serious. When the small black lamb jumped from the truck, she landed on her knees. We assumed it was just the height of the truck bed. Our old Ford one-ton was tall, and perhaps the ramp looked intimidating. We were wrong. She wasn't clumsy or nervous; she was sick. Her knees buckled because her hooves were tender with footrot.

She couldn't stand for long, but she could still spread infection. Without a proper isolation pen, we tethered her in a grassy area with clean soil and plenty of space.

She was miserable. Sheep are social creatures, and she missed her flock. Her lonely bleats tugged at my heart, so I spent hours beside her, learning where she liked to be scratched, rubbing her sweet face, and letting her know she wasn't forgotten.

Gradually, her hooves healed, and—miraculously—none of our other sheep were infected. Angie, for her part, lived out her days with us and died of old age, still loved and doted on.

In purely financial terms, both animals were losses, though charming. Many of our farming decisions were like that: heart first, balance sheet second.

Chapter Twenty-Seven
THE SHEEP MIDWIFE

Bottom line be damned. We had animals to save, which is how I found myself one day with my arm up a sheep's bum.

We had a sheepdog—sweet-natured, loyal, and not quite blessed with the breed's legendary intelligence. All we needed now were some sheep.

A friend offered to sell his small flock of Dorsets and gave us what he called a "can't-miss" price. Suddenly, we were the proud owners of a ram and several pregnant ewes.

We plunged into study mode, devouring books on sheep nutrition, veterinary care, and anything that might help us become competent shepherds. As the weeks passed, ewe bellies swelled, and it became clear we'd be lambing in late fall and early winter—hardly ideal for beginners.

One Friday night, one of the ewes went into labour. We had read enough to know that most births took care of themselves, so we waited, expecting nature to do what nature does best. But something wasn't right. The ewe strained and bleated, eyes wide with effort. My husband could see she was in trouble.

An after-hours vet call would cost a fortune. I was the squeamish one, but I also had the smaller arms. That settled it. I went to the house to scrub, grabbed our well-thumbed guide, *Raising Sheep the Modern Way,* and returned to the barn praying the lamb would have arrived by the time I got back.

She had not. The ewe was exhausted and barely pushing. Richard opened the book to the chapter on difficult deliveries and began reading the text. I knelt behind the ewe, heart hammering, and carefully slid my arm into a place no arm of mine had ever been. It was wet, warm, and utterly foreign.

As Richard described what I should be encountering, I tried to envision what I was touching. I thought of those pots of peeled grapes and boiled spaghetti that passed for eyeballs and guts at the Halloween parties of my childhood. Only, this was real, and two lives were at stake. Moving carefully so as not to damage the uterus, I felt for the front hooves and nose that should have been aimed at the exit. Even for someone as unfamiliar with a ewe's interior as I was, I knew what I was touching was the bum end of a lamb. My first midwifery experience was going to be a breech birth.

I made a few failed attempts to reposition the little one, but I was too inexperienced. By now, the ewe and I were both stressed. If I did not act quickly, both mother and lamb could die. I stopped trying to turn the lamb and pondered how to perform my first, difficult lamb delivery.

I found one tiny hind leg and gently pulled it straight. Holding the hoof with my free hand, I reached in again and straightened the second leg. With both hooves lined up, I held on for dear life while I firmly and steadily pulled.

The ewe cooperated, bearing down as I tugged. When I finally freed the limp lamb, slick with afterbirth, I was sure the little one was dead. We cleared her nose and mouth and placed her by her mother's head.

The ewe, spent as she was, lifted her head and began to lick the little one clean, murmuring softly. After a few moments, the lamb stirred, shook her head, and gave a faint bleat.

She was alive.

I sank back in the hay, overwhelmed by relief and wonder.

The breech lamb survived and was soon leaping about the pasture. Many lambs followed in later seasons, and we assisted with more difficult births, but none ever equaled that first one, when a pair of newly minted sheep farmers learned what it meant to help bring life into the world.

Chapter Twenty-Eight
TOO MUCH OF A GOOD THING

Gradually, we acquired more sheep. The animals knew what they needed. Two neophyte farmers had to learn. We turned to friends in the smallholder community, but much of our learning came from constantly reading books and magazines, from trial and error, and from the animals themselves. When we made mistakes, they paid the price. I knew we were trying our best, but I was hard on myself for not always getting it right.

One of our first Dorset ewes to lamb gave birth to healthy twins. We named them Thanks and Giving and watched them grow into fat, joyous creatures. Two months into motherhood, the ewe looked thin, so we gave her extra grain. She devoured it eagerly, but her system had no time to adjust. The sudden change upset the balance in her rumen, increasing a bacterium called *Streptococcus bovis*. That led to an overload of lactic acid, followed by diarrhea and bloat.

By the time our vet arrived, we were sure she was dying. He gave her penicillin and sent me to the house for baking soda and water, which he poured down her throat. For the next two days, we took turns nursing her. Richard administered the penicillin. I dosed her with soda water and fed milk replacement to her twins. For a while, she seemed to rally.

By mid-afternoon on the third day, she was failing fast. The vet told me to try walking her, to let her lungs drain. She managed for ten minutes, then collapsed. By the time he arrived to try cortisone and a different antibiotic, she was too weak to stand.

Three days after we had tried to do her a kindness with extra grain, she died. We had taken a break to go out for a quick dinner. She was gone when we returned. Her lambs had already weaned themselves. They did not need bottle feeding, but I could not look at them without tears. Causing suffering was bad enough. Causing death broke my heart.

With her death, we lost more than the cost of the ewe and the vet. We lost a friend. She had a habit of hanging around the barn after the others headed out to pasture, hoping for an extra apple before joining them. She was gentle and attentive, a devoted mother. Even as she grew weaker, her lambs stayed by her side, leaving only to graze or drink.

I was the last to check on her the night she died. Tears streamed down my face as I sat with her to say goodbye. I thanked her for her lambs and her friendliness and asked forgiveness for what we had done to bring on her death. Then I went back to the house to tell Richard. He and his younger son dug a grave for her in the barnyard. They lined it with soft hay and placed her body inside. It was a small gesture, but one that acknowledged her life had meaning.

After her death, the lambs attached themselves to a ewe with triplets. The triplets were slightly younger and still nursing. When they called for their mother, the orphans called too. When no one answered, they looked around, bewildered, then went back to grazing or followed the others.

Our growing menagerie revealed the complex emotional lives of animals. New lambs gamboling in the field were the very definition of joy. Cows bawling for days over a stillborn calf expressed a grief as deep as any I had known. A restless camel swaying for hours embodied frustration and stress. Each animal was distinct—an individual with moods, quirks, and feelings.

Fortunately, researchers have begun to challenge old notions about animal behaviour. Science is finally confirming what farmers, pet owners, and animal lovers have long known: sentient beings live among us, and we are their stewards.

Chapter Twenty-Nine
AUCTION IS A SEVEN-LETTER WORD

The vision of eggs, bacon, and toast washed down with coffee lured us out for an early breakfast. We were auction-bound, eager for the game. I was usually the foot-dragger, but I could be swept up in the fun of it. Weeks or even years later, Richard would reach into a bucket of miscellaneous tools and odd bits we'd picked up at an auction, pull out exactly the piece he needed, and grin. "Seven-letter word!" he'd say, his shorthand for *auction treasure*.

One Saturday, we bought two Finnsheep—a brown ewe and her darker lamb. They loaded easily into the truck, but the troubles began once we reached the farm. When we tried to check the ewe's hooves, she broke through a fence and tangled herself in barbed wire. I managed to free her and guide her into a pasture, but one of our rams took immediate interest and chased her until she bolted through another fence. Her sprinting made us think she must be healthy enough. We patched the fences and decided to let her settle in overnight.

By morning, she was feeding on her knees and lying down often. When we finally managed to pen her, we saw that her hooves had nearly rotted away. She must have been in terrible pain for a long time before her owner sent her to the auction. There was no saving her. Footrot is highly contagious, and her lamb was almost certainly infected. With heavy hearts, we sent them both to the butcher.

We fenced off a clean section of pasture and confined the rest of the flock there. We scrubbed and disinfected troughs, bedding, and pens, then built a footbath filled with medicinal solution. Each sheep had to walk through it on the way between barn and pasture.

The work paid off. None of the others became infected. The two beautiful Finnsheep were not so lucky. They ended up in our freezer—an outcome that still carried sadness, but also relief that their suffering was over.

And the next Saturday, we were back at the auction.

Chapter Thirty
BEWARE THE BATTERING RAM

A sturdy, vigorous ram was one of our first male sheep. Big Al should have come with a warning sign. One sunny day, he stood by the fence and nuzzled my hand—all friendliness and calm. That lasted until I entered the field where our Dorset ram was to do his important work of providing us with sturdy lambs.

I was alone. Assured Big Al meant no harm, I started checking the ewes. I had barely turned my back when something slammed into me, sending me headfirst into the grass.

My bum smarting from the ram's battering, I lifted my head and looked around for Al. He was grazing peacefully nearby, not even glancing my way. He kept munching grass until I rose to my hands and knees. If a sheep's eyes can light up, Al's were on high beams. If a sheep has a sense of humour, Al was laughing uproariously. Fixing me with his stare, he slowly lowered his head.

I flattened into the grass. He went back to grazing. I tried the hands-and-knees pose again. He grinned and resumed his battering stance. I dropped back down.

My husband would not be home for hours. If Al crushed my spine or broke my hip, I would lie in the field, unable to move.

Mobile phones were not yet ubiquitous, so I would not be able to call for help.

In a maneuver reminiscent of soldiers inching their way across a field, bullets whizzing overhead, I elbowed and dragged my way toward the fence. Al quietly moved with me, never looking straight at me, always keeping me in easy battering range. The moment I touched the fence, I shot to my feet and scrambled over it. Al charged but stopped short when he saw he had been outwitted.

He lowered his head and tore a mouthful of grass, pretending that had been his plan all along. But in his heart of hearts, Al knew he had won. He ruled the field. I could never again turn my back on him—nor enter his domain without running shoes.

Big Al was my first battering ram, but he was not, alas, the last.

Chapter Thirty-One
NOT QUITE ENOUGH COURAGE

Our next battering ram was Jake. He was an auction ram—chocolate brown, likely a Jacob cross—and he had attitude. We put him in a temporary stall to ensure he was healthy before joining the other rams. The old barn's stalls were fenced with sturdy 2x6 boards. Surely, we thought, nothing could break through that.

Jake proved us wrong within minutes. He splintered the top boards of the gate with his massive horns. If he could have backed up further, he might have reduced the entire stall to kindling.

Richard hurriedly bought stronger lumber and hammered it in place. The gate rebuilt, the pen reinforced, Jake's health—or at least his strength—well established, we decided to introduce him to the other rams.

We built a temporary run through the barn. I would keep Al and Callie (our sweet-natured Jacob ram) in their stall. Richard and Phantom, our eight-month-old Border Collie, would push Jake my way. I would jump aside, letting him run past me into the ram stall.

Remembering my encounter with Al, I chickened out. Instead of standing in the gate, I climbed the fence and reached down to swing it open. Self-preservation trumped my husband's instructions. From my perch, I waved my arms uselessly as both rams ignored me and surged past.

Richard was about to be caught in the path of all three rams when Phantom stopped hesitating and took charge. Barking and snapping, he drove the charging trio into the big stall. We slammed and barred the gate, then stood back, hearts pounding, to watch them sort it out. It was a rare moment of glory for Phantom, who until that day had shown little interest in sheep.

With the rams now in one pen, we watched to see how they would interact. True to his gentle nature, Callie sniffed Jake but did not bother to challenge him. Big Al, terror of the barnyard and giant of my imagination, met Jake head-on a couple of times but without conviction. In less than a minute, Jake let the two rams know they were dealing with a superior animal.

That was the end of it. They settled down and grazed together peacefully when we let them out into the ram pasture—at least until the ewes were in heat. The alluring scent of potential mates sent them into frenzies of bashing until they'd had a chance to exercise their hormones.

With the three rams coexisting in harmony, the ewes sweetly grazing, life with sheep began to feel almost routine. We were beginning to get the hang of this farming thing.

Chapter Thirty-Two

ANIMALS DO TALK ON CHRISTMAS EVE

Our menagerie had grown to the point that we could no longer be away overnight. As Christmas neared, a family celebration was out of the question unless people came to us. When Richard's siblings agreed to come from Vancouver to spend the holiday with us on Vancouver Island, we were thrilled.

On Christmas Eve we feasted, talked, told stories, and laughed. Probably fuelled by one too many festive libations, I suggested we head down to the barn at midnight.

I vaguely remembered an old legend about animals talking on that holy night. They had comforted the Baby Jesus as he lay in the manger, so God gave them the power of human speech. It was one brief moment of magic. With a barn full of animals, I remembered the story and wondered.

The world often seemed enchanted. I had never been near a barn on Christmas Eve. Perhaps our friendly beasts would talk.

I persuaded a group of skeptical adults to troop down to the barn at midnight. What happened was, sweet and unexpected. A few years later, when Richard and I began writing and performing cowboy poetry, I turned the story into a poem. *American Cowboy* published it in 1994.

Stock Talk Christmas Eve

One wintry night the relatives
Were gathered in our barn.
They'd all come from their city homes
For Christmas at the farm.

'Twas Christmas Eve, and just before
The wassail was passed 'round,
We donned our coats and headed down
To hear the magic sound

Of animals at midnight,
For then the power of speech
Is given to all sheep and cows,
Or so I'd heard it preached.

My husband, he was skeptical,
The relatives amused.
They figured I'd gone round the bend
Since donning country shoes.

But to the barn they gamely trooped.
They'd humour me this time.
We flipped the switch and walked into
A scene that was sublime.

The sheep were calmly bedded down.
They looked, then turned away,
For we'd disturbed their peaceful rest
And hadn't brought them hay.

I thought of tales of talking beasts.
"Let's sing to them!" I cried.
Embarrassed silence met my plea.
"Let's not," my husband sighed.

No word came from those woolly heads.
I blushed and murmured low,
"They prob'ly talk when we're not here.
I guess we'd better go."

Then coming from a darkened stall,
We heard a little cry,
Soon followed by a throaty one
That pulled us to draw nigh

And watch a newborn struggle up
To reach her mother's teat.
She crumpled, rose, and tried again
On tiny cloven feet.

While ewe and lamb crooned soft and low,
We cleared our throats and sang
Of friendly beasts and silent nights
And bells that angels rang.

Then all the livestock in the barn
Began to bleat and crow
And oink and quack and gobble
In the languages they know.

The relatives fell silent
Till one softly observed,
"That's the closest thing to talking
This city dude has heard."

*So maybe friendly beasts don't speak
In English or Chinese,
But if you listen close
You'll hear them talk on Christmas Eve.*

Chapter Thirty-Three

LESSONS FROM THE SHEEP BARN

The barn's magic on Christmas Eve lingered in my mind long after the last carol faded and the animals returned to their usual rhythms. Yet as winter deepened, the enchantment of soft hay and twinkling lights gave way to the real, relentless work of caring for a growing flock and experiencing our first sheep births.

We had bought a small flock of pregnant Dorsets from friends. Through the coldest months of our first winter on the farm, the ewes were popping out lambs. Most of the little ones thrived, but we discovered our grass lacked selenium.

Selenium deficiency leads to white muscle disease. Lambs who suffer from it may hunch their backs, walk stiffly, and become too weak to nurse. Learning how to prevent and deal with it means vet bills, selenium shots, enriched feed, and a fair share of worry, but it also leads to healthy lambs.

We learned to watch for ewes separating themselves in preparation for giving birth and to judge how long to wait before intervening. Heat lamps were our friends during cold spells, keeping birth-wet lambs warm while their mothers licked them dry. I could not resist nuzzling those adorable lambs. After some unpleasant butting, I learned that cuddling baby rams could backfire if I did not teach them that practicing on humans was off limits.

Sometimes we tried too hard to save ailing animals, refusing to accept nature's course. One Jacob ewe, Curly, was already failing when she gave birth to a premature lamb. She could not stand on her own. The vet warned she had nerve damage and would not recover. My husband rigged up a sling, and together we tried to teach her to walk again. She lingered for weeks, sometimes staggering like a drunken sailor but mostly not walking at all.

On a warm day in March, she seemed to improve. We opened her pen and let her outside. Overjoyed, she ran toward the other sheep—but collapsed in a heap. My husband looked like a painting of the Good Shepherd as he carried her back to the barn. She never walked again. Our well-intentioned physiotherapy had only prolonged her suffering.

About the same time, a Southdown ewe developed pregnancy toxemia and lost her fleece. We feared she would be another casualty. We brought the pink, hairless ewe into the barn, tucked her into a pen with one of our ubiquitous heat lamps, and hoped for the best. Defying my endless worries, she recovered completely.

Then there was Topknot, a Dorset ewe who gave birth to an 11-pound lamb, all on her own, with astonishing ease. Spats, a Suffolk ewe who reliably gave us triplets, had delivered healthy triplets a day or so before. We tried bonding Spats's smallest lamb to Topknot, hoping her abundant milk would nourish him. We even rubbed the afterbirth from her own lamb onto him—but Topknot was not fooled. Each time we brought the tiny ram to her teats, she butted him firmly away.

We thought if we could keep him feeding from her long enough for the milk to work its way through his system, he would smell familiar to her. We overlooked the bonding that took place before he ever left her womb. When he popped into the world, he knew his mother's voice. And she knew his. We gave the little guy back to Spats, who welcomed him easily. To help him catch up to his siblings, we occasionally bottle-fed him. Spats had plenty of milk, but the little ram loved the extra food and, most of all, the extra cuddles.

They lived. They died. I laughed. I wept. Most of our lambs were healthy, born live, and thriving—but I never lost the sense that these beautiful creatures in our care deserved more experienced hands than I could offer.

Chapter Thirty-Four
OVINE ESCAPE ARTIST

The two lambs were born within days of each other. Both were heritage breeds—species dwindling in numbers but kept alive by people like us, determined to maintain biodiversity.

One little ram was a Southdown. With their teddy-bear ears and short legs, Southdowns look like four-hooved Ewoks.

The other was a Jacob, with its signature piebald fleece—a mottled black and white pattern characteristic of the breed. He had the nubs of two future horns, though Jacobs can have as many as six. They may be descended from the piebald sheep specially bred by Jacob, son of Laban, in the biblical story.

Before the rams matured enough to be interested in the ewes, all the lambs shared the same pasture and pens with their mothers. Within days of birth, they were leaping, jumping, and racing around the field—a moving sea of joyous life.

Friendship chemistry is as mysterious among sheep as it is among humans. The Jacob and Southdown ram lambs recognized each other as kindred spirits. They were barely beyond the tottering stage when they became inseparable. Unless they were nursing, they were together.

Rams can become sexually active as early as four months. To prevent them from disrupting our breeding plans, we separated the growing lambs into breed-specific pastures. The Dorsets were in one, the Southdowns in another, the Jacobs in yet another. The sorting went relatively easily. We went to bed satisfied that there would be no mixed-breed accidents among the next round of lambs.

In the morning, we found the young Southdown ram in the Jacob pasture. Several sturdy fences separated him from his friend, but he had found a way to circumvent them all. He and his pal were grazing peacefully, side by side, as they had always done.

We checked fences, gates, and posts. Nothing appeared broken, bent, or loose. Undeterred, we tried again. This time, the Southdown was aware of our intentions and even harder to separate. All day, their mournful bleating punctuated our chores. The next morning, the Southdown was back with his pal—still with no sign of how he had managed the escape.

After two more attempts, we gave up. Their friendship was stronger than our will to separate them. Eventually, they matured enough to spend brief periods apart to impregnate a few ewes—but they remained willing to be apart only when the sweet scent of ewes in heat set their upper lips curling back with curiosity.

I had never thought much about friendship between species other than humans. Except when rams were butting heads over breeding rights, they all seemed to get along peacefully enough. I didn't understand the depth of affection they could hold for each other until these two refused to be separated.

Chapter Thirty-Five

"IF YOU'VE GOT LIVESTOCK, YOU'VE GOT DEAD STOCK"

When our first auction chicken died, I wept. She was a Rhode Island Red, one of my cheerful gardening companions. She was undoubtedly a "spent hen"—too old to lay eggs—when we bought her. Someone reluctant to consign her to the stew pot had dropped her off, hoping to earn a few pennies for his trouble.

She was never a pet, but I liked seeing her wander the yard. She was already stiff when I found her body. A short while later, a "real" farmer dropped by. She looked at my reddened eyes, saw the dead chicken, shook her head, and said, "If you've got livestock, you've got deadstock."

Until we gave farming a try, I had little experience of death. When I was ten, my Grandma Matthews died. She was a grandmother by love rather than by blood. She and her husband lived next to Aunt Grace and Uncle Dewey. The two families spent a lot of time looking after my brother and me while our single mother worked five and a half days a week as a bookkeeper. When Grandma Matthews died, Mother decided my brother and I were old enough to face the reality of death. She took us to the funeral home to view the body. I had nightmares for years.

That was the last dead body I saw until, three and a half decades later, I began farming. When the old hen died, I was crushed.

She was just the beginning. Tiny chicks drowned, froze, or were smothered. Lambs slid lifeless into the world. Cows bellowed their grief as they licked stillborn calves. A pig succumbed to a savage dog attack. The deaths came in a long string, from our first days of farming to our last—and that does not include the creatures whose lives we deliberately cut short to sustain the farm.

Death was a regular reminder that livestock farming was not in my blood, despite all the joy it gave me. Every creature's life is finite. Some end before birth, others after injury or illness, yet others when old age finally shuts things down. Birth and death are natural bookends around the stories of our lives. Still, I never became peaceful with it. I mourned each death. I hated each trip to the slaughterhouse, the old blue truck laden with animals who had viewed us as benign companions until then.

And yet, when the flesh of a beloved pig or sweet lamb graced our table, I felt enormous gratitude for the animal whose ears I had scratched, whose belly I had rubbed.

The animals whose lives blessed my years as a reluctant farmer taught me lessons that marked me. I still eat meat, though very little. I am cautious about my suppliers. I want to know that the animals whose flesh becomes part of my body were respected and loved before their lives were cut short.

When animals are part of a well-managed, mixed farm, they contribute to soil fertility. Unfortunately, many are cogs in the wheels that run factory farms. They are sentient beings forced into prison-like conditions. They suffer. Millions are tossed on the garbage heap when food-safety lapses lead to massive meat recalls. They contribute heavily to environmental degradation and climate change because of the crowded conditions in which they live their short lives.

Somewhere in the mists of history, certain animals formed uneasy alliances with their human neighbours. They traded freedom and vulnerability for food and protection. Livestock have kept their side of the bargain. We need to keep ours.

Chapter Thirty-Six
SPRINKLE WITH HYPOCRISY, & SERVE HOT

Hostility radiated from her eyes as she looked at the flyer I had handed her.

"You sell lamb and chicken," she said, a note of shock in her voice. "I will never buy anything from you."

Richard and I were at our first craft fair, enthusiastically sharing our passion for heritage breeds and seeds. We had been having a friendly conversation with the woman. We must have been talking seeds, because she was rapt. She looked over the jars of home-canned salsas and jams, studied the dried-herb arrangements, and asked thoughtful questions.

We were fishing with the right bait, slowly reeling her in to buy some of our offerings. That's when I made a tactical error: I handed her an order form.

Her expression changed instantly. Images of lambs and chickens marching off to slaughter seemed to flash across her mind. We were no longer friendly advocates for sustainable farming; we were monsters of the worst order. She flung down the form and strode away, leaving us blinking in surprise.

As she left, I noticed her outfit: a soft wool tunic, cinched with a leather belt. Leather shoes completed the ensemble. The irony was not lost on me. Wool and leather come from animals too—sometimes under conditions far removed from idyllic farms. It reminded me that everyone makes compromises, some deliberate, some unconscious.

I understood her dismay, which mirrored some of my own discomfort with livestock farming. She reminded me to stay aware of the compromises I make. I rail against harmful farming practices and look for food that is grown or produced sustainably—but not all the time. Too often, I overlook other compromises entirely, like the environmental cost of single-use plastics, chemically treated foods, or even the clothing I buy secondhand without checking its content or origins.

More than thirty years later, the incident still comes to mind. It reminds me to acknowledge my personal choices before judging others, and to accept that perfection is not in my future. What matters is striving to live responsibly, with humility, and with a willingness to keep learning.

Maybe that's why I had a soft spot for the weak and ailing animals on our farm. Whatever guilt or doubt I carried found relief in tending to their needs.

Chapter Thirty-Seven
MOTHERING THE WEAK ONES

As a childless woman, I had no mothering experience. Whatever parenting instincts survived my earlier years of infertility testing and a crumbling marriage came to the fore as I watched eggs hatch and lambs enter the world.

I felt connected to the young life pecking, peeping, baaing, and leaping around the farm. But it was the weak ones who stirred the deepest tenderness in me.

As a newly minted farmer, I held the naïve notion that all our animals would be healthy. We farmed organically and raised our livestock on grass, without antibiotics or growth-inducing chemicals. We read constantly and sought advice from experienced farmers and agricultural experts, eager to give our livestock the best care. Surely, I thought, they would all thrive.

I hadn't accounted for the fragility of life, no matter how carefully we tend it. Some chicks emerge from their shells with weak legs or bad hearts. Lambs die in the womb; ewes in birthing. An otherwise healthy lamb may be stepped on, attacked by predators, or rejected by her mother. The list of possible injuries, diseases, or genetic abnormalities is endless.

One January night, a hugely pregnant ewe gave birth to a scrawny little ram lamb. I found him during my final check of the animals. He lay in the hay, weak and listless. I ran back to the house for help. Richard held the ewe while I tried to guide the little one to her teat. He was too weak to suckle.

We carried him to the house, where Richard prepared a bottle of special formula for baby lambs. I held him against my skin to warm his little body. Gradually, he found the strength to nurse, then slept soundly in a towel-lined box beside our bed. He woke at dawn, drank eagerly, and drifted back to sleep.

By morning, he was strong enough to toddle about on wobbly legs, keeping us busy cleaning up after him. When Richard took him back to his mother in the barn, we discovered that a stretch of sub-zero weather had frozen the pipes. While he worked to thaw them, I hauled buckets of water from the house for a menagerie that now included fifty sheep, four goats, four pigs, five geese, sixteen Muscovy ducks, five mallards, and a hundred and fifty chickens.

When I checked on the little lamb, I found he was still not nursing from his mother. He tottered over and nudged my leg for milk. The ewe was patient when I tried holding him to her teats, but the lamb was still too weak to suckle.

Richard took him to the vet and came home with a round of penicillin. Born on one of the coldest nights of the year, the newborn had contracted pneumonia. If he survived, he would eventually end up in our freezer. We guaranteed our customers meat from animals that had not been medicated.

So he became a house lamb, gaining weight on bottled lamb milk replacer and regular feedings from his mother. We buttoned him into baby sweaters for his trips to the cold barn. His mother always called to him, always welcomed him back.

As he gained strength, he began to prance and leap like a normal lamb. I called him *Squirt* because he peed everywhere. When he seemed sturdy enough to handle the chill and could nurse on his own, we left him in the barn with his mother.

The little guy thrived, though he was always on the small side. He viewed us as family, running for a scratch when he heard our voices. We almost felt invincible after saving him, but we learned repeatedly that we were not—and that sometimes our interventions only prolonged suffering.

We never forgot that lesson. It returned with painful clarity years later, on another farm, when we tried to save a goat kid rejected by his mother.

Chapter Thirty-Eight
ISLAND INTERLUDE

The contained size of Vancouver Island suited me. City, country, open spaces, high mountains—Vancouver Island had it all. From Auchinachie Farm in Duncan, we often drove the backroads past small farms and rural acreages. Sometimes we headed north to Chemainus, wandering among shops filled with local artisans' work and pausing to admire the murals that drew deep-pocketed tourists to a town abandoned by forestry. When a theatre opened there, we sometimes bought tickets and took in a live show.

Other days, we drove south to Cowichan Bay for fresh seafood and water views. We followed winding roads connecting Cobble Hill and Mill Bay, often stopping by Fairburn Farm to visit Darrel and Anthea Archer.

Vancouver Island was dotted with small farms offering artisanal foods. I had been a foodie long before buying a farm—shopping at organic co-ops in Rochester and Seattle, refilling jars at bulk stores, and supporting farmers' markets. When we helped start an organization for people with small acreages and big dreams, we found kindred spirits. Many were professionals who had left high-paying jobs in search of a simpler, more grounded life. We met to share ideas, learn about marketing and processing, and commiserate over the unpredictability of weather, animals, and markets. Most of us were new to farm life, slowly learning that land and livestock come with no instruction manuals or guarantees.

Just as I had learned to watch the moods and needs of each animal, I found myself reading the rhythms of the island—the seasons, the tides, the personalities of fellow smallholders. Nourishing friendships grew from those gatherings, much like the way I had nurtured lambs and chicks. Both required patience, attention, and a willingness to learn from small signals, to celebrate small victories, and to accept setbacks with grace.

The long, slow ferry rides to and from the island became a kind of meditation. We travelled through a seascape dotted with smaller islands, visiting family on the mainland, dining with friends on Thetis, or wandering Salt Spring before returning home. Once on the ferry, I could let go of deadlines and demands. I read, sipped coffee, watched the water, and often ran into island friends I hadn't seen in a while.

Vancouver Island had become home. Its beauty and mild climate drew people from across Canada and beyond, many of them bringing creativity and determination to their new lives.

At the Venturi-Schulze Vineyards, we sampled balsamic vinegar made from a "mother" given to Giordano at birth. He and Marilyn Schulze tended their vines without pesticides or additives, honouring old-world methods. When a spoonful of their 25-year-old vinegar touched my tongue, I understood the difference between artisanal care and industrial convenience—the same devotion to life and attention to detail that we tried to give to our sheep and chickens.

Then there was Merridale Cidery, whose superbly fermented juice slid down easily on hot festival days. Their business grew, but their commitment to quality remained constant. At Cowichan Bay Farm, we shared visits and farming philosophies with Lyle and Fiona Young. Lyle, a fourth-generation island farmer, held fast to humane and organic practices. Like so many smallholders producing exceptional food, they both worked off-farm jobs to support their dream.

Fairburn Farm was always a warm destination, its leafy drive leading to the welcome of Darrel and Anthea Archer. Back then they were running a bed and breakfast alongside their mixed farm, long before they brought water buffalo to the island. When BSE was discovered in a Danish cow, the government ordered the slaughter of their imported herd. Tests later showed not a trace of the disease. The Archers were heartbroken, but their dream endured. Years later, their son and his wife carried the torch, producing buffalo mozzarella in Quebec.

Nick Usborne, still juggling his marketing career, was establishing a heritage-breeds farm, hoping to inspire visitors with animals rarely seen anymore.

I could have stayed on that island forever. It offered beauty, community, and a rhythm of life that fed my spirit. Yet just as I had learned that life on the farm demanded adaptation, I understood that our journey, too, required change. Richard's heart was elsewhere, drawn north to the gold rush country of his imagination: the Cariboo. And so we packed our dreams, our animals, and our lessons in care, moving toward the next adventure with gratitude for all we had learned.

Chapter Thirty-Nine
A DREAM STRONG AS THE LURE OF GOLD

From the time I first set foot in a mobile home on the edge of Duncan to the day Richard proposed we look at property in British Columbia's heartland, nearly four years had passed. Our farm was beginning to show a modest profit, and we had become well established on Vancouver Island. The beautiful isle felt like home to me.

We had accomplished a great deal during our time there. We launched a heritage breeds and seeds program on our farm, co-founded an organization for the valley's small farms, and helped start a farmers' market. A few of us who owned farms with long histories organized a heritage farm tour. We often got together with other smallholders to discuss ways to strengthen opportunities for local food producers in the Cowichan Valley.

We also performed our unique blend of music and storytelling, entertaining audiences from California to the Yukon, Toronto to Vancouver. A contract with the Cowichan Folk Guild as artistic directors connected us with remarkable musicians from across Canada and beyond. I built a steady demand for my storytelling courses through the island's continuing education programs, and with friends' help, we converted our barn loft into a performance space for house concerts. Among the musicians, we welcomed storytellers like Liz Weir and Billy Teare from Northern Ireland and Vi Hilbert from Washington State—each one adding a new thread to the tapestry of community we had woven on the island.

When we needed a break from farm work, we escaped to Victoria for days of research in the provincial archives. Thanks to Richard's years of writing about B.C. history, he had close ties with the archivists, who always went out of their way to assist us. I loved those city interludes—digging for historical treasures, then walking the waterfront or lingering over dinner in small restaurants, feeling fully alive in both worlds: the quiet farm and the bustling city.

Richard was steeped in the lore of the Cariboo gold rush and the dreamers who had followed the Fraser River north in search of fortune. Adventurers had scoured the riverbanks during the 1858 gold rush, and when William "Billy" Barker struck the richest vein of gold along Williams Creek in 1862, the real rush began. Richard had written extensively about the Cariboo and spent summers as a street interpreter in the restored town of Barkerville.

A friend once told me Richard was born in the wrong century. His heart belonged to those rugged souls who carved wagon roads into the mountainous interior.

For me, our island home—with its lush fields, mild climate, and proximity to the sea—was ideal. For Richard, the call of the Cariboo was irresistible. We submitted a bid to manage Cottonwood House, a historic site on the Barkerville Highway, but the contract went to another applicant.

When he learned a historic ranch was for sale, his imagination leapt north again. The property had once been a stopping place on the trail to Harper's Camp (now Horsefly), where the first gold of the Cariboo rush was panned in 1859. To him, it was more than a ranch; it was a piece of living history. I could see the spark in his eyes and didn't want to dim it with my hesitation. So we arranged for someone to care for our animals, packed the car, and set out in the dead of winter to see the place that had captured his heart long before it captured mine.

Chapter Forty
A DAUNTING INTRODUCTION

In February 1994, the Cariboo lay deep in winter. From the top of the long, snow-packed driveway leading into Pioneer Ranch, we looked down on a landscape of pristine white fields punctuated by dark clusters of conifers—and, at its center, a house and outbuildings badly in need of attention.

Inside, the rooms were small and dim. One bedroom could only be reached by walking through another. A brick fireplace dominated the living room, turning it into a long, narrow tunnel.

Water came from a pipe connected to the neighbours' well. They had been renting the property and allowed us temporary access, but whoever bought Pioneer Ranch would have to dig a new well within a year.

Between the house and the pastures beyond stood a tiny, two-story cabin—the original home from the 1800s. Nearby was a log chicken coop, sturdier than any chickens required but uninsulated, offering little protection against the Cariboo cold. Between the coop and the house stood another log structure that might serve as tool storage.

Old frame buildings flanked the ranch house. On one side, a

garage offered space for a workshop and storage. On the other stood the relic of what had once been the stage stop. Later, we found a list of goods and prices penciled on one of its beams, probably by someone with no paper at hand. Holes in the roof let in snow and daylight. They would have to be patched before the building could be put to use.

To the left of the coop and shed was a small structure about the size of a miner's cabin. It would become the first building renovated, a modest home for Richard's son, who shared his father's Cariboo dreams and had offered to lend a hand.

Beyond that were two log barns. The larger, in decent shape, would shelter birthing or ailing animals. Its roomy hayloft could hold a winter's supply. The smaller barn was divided into two rooms: one for sheep to weather storms, the other destined to store meat freezers and boxes of books from our publishing house, Winter Quarters Press.

As we wandered through the snow, I saw isolation and a mountain of work. Richard saw history and possibility. He was alive with ideas and energy, eager to begin. I was silent, taking in the snow, the cold, the weight of what this move would mean.

I could not yet imagine learning to love the ranch or this vast, demanding country. But love has a way of catching us unaware—sometimes only after we've said yes to something that terrifies us. The introduction was daunting. The adventure had already begun.

Chapter Forty-One
MA & PA HEAD NORTH

With equal parts excitement and uncertainty, we signed the papers to buy the property and returned to our island home. Not long after our trip to see the ranch, we flew to Toronto to perform at a storytelling festival. While we were there, the news came that our offer had been accepted. I took a deep breath and smiled, trying to absorb the enormity of what we had set in motion.

To friends and colleagues, our decision was a mystery. We were leaving an idyllic home on Vancouver Island for a remote patch of land far to the north—a leap into a new and uncharted life. To me, it felt both daring and bewildering, like stepping off a familiar map.

We put Auchinachie Farm up for sale and did not have long to wait. The market was strong, and our friend Richard Hughes, a skilled sales agent, got us the best possible price.

We threw ourselves into packing, clearing out, and leaving the farm for its next owners. I knew there would be rough patches ahead. What I did not foresee was that I would fall in love with the landscape, adore the animals in my care, grow strong mending fences and chucking bales, make friends for life, watch bear cubs romp in the fields, and start a successful consulting career. I would also learn to redefine "remote" to exclude any place with a paved road, telephones, and electricity.

All that lay in the hazy, unknown future. In those final weeks on Vancouver Island, as we said our goodbyes, I counted the miles to the Cariboo and did what I do best: I worried.

We made the move in stages. We decided to renovate the ranch house before moving in, so we packed everything that could be stored in the ranch's garage. Richard's son and one of his brothers made several round trips with Old Blue, our 1967 bright-blue half-ton Ford truck. They unloaded at Pioneer Ranch and made the nine-hour drive back to the island. The last trip would be with the animals who had shared our farm life.

As we said our farewells, I assured friends I was excited about the move. Inside, I was dreading every bit of it.

And then we were gone, looking back one last time at the double hump of Mount Prevost, the spiral mandala of the herb garden, the porch that wrapped around the house, the pastures we had fenced, the flower and vegetable gardens, the cherry and apple trees, and the neighbouring farms that had framed our days.

With all the squawking, clucking, oinking, and baaing coming from the back of Old Blue, we looked like the Clampett family from *The Beverly Hillbillies*. Only they struck gold and headed for the city. We moved in the opposite direction—toward a new frontier, not of fortune, but of faith in our own resilience.

Auchinachie Farm from the porch, with the double humps of Mount Prevost in the distance. Photo taken by one of our WWOOFers.

The farm in winter, taken from the fields. Photo by Cathryn

Turkey Baby and her foster mother, Millie, on their grazing rounds. Photo by Cathryn

Cathryn's first experience of sheep midwifery. Our renter took a series of photos of the experience.

Thanks to Richard's connections with local newspapers and Cathryn's computer skills, we were able to gain a lot of publicity for our farming activities.

The WWOOFer who took the photograph of the two of us and of our sweet Jacob ram, Callie, sent us a calendar with farm photos.

It's obvious why Southdown sheep looked like woolly Ewoks to us. Photo by Cathryn

Part Two

PIONEER RANCH, CARIBOO, 1994-2004

Chapter Forty-Two
ARRIVING AT THE RANCH

A colourful caravan boarded the ferry. Old Blue was packed with sheep, pigs, ducks, geese, and chickens. Every corner of our two cars was filled. Driving onto the ferry marked a milestone: we would no longer be residents of the green isle.

We were already expecting our first visitors. Richard's father and his partner were driving up from Vancouver to see Pioneer Ranch and help us settle in.

I loaded my little Honda Civic, tucking in a crate for our orange tabby cat, Jalebi. He had come to us when his previous owner married someone with cat allergies. The ideal barn cat, he was content to live outside. Riding in a car was far from his idea of fun.

We set out together, boarding the same ferry. Once on the mainland, I lingered along the way, stopping now and then to stretch and gather my thoughts. The idea of turning back crossed my mind, but the sheer logistics of uprooting ourselves loomed larger than the unknown ahead.

By the time I finally drove down the ranch driveway—several hours after the others had arrived—Richard was understandably anxious. It was June 1994, before cell phones, and he had no way of knowing where I was. Aware of my mixed feelings about the move, he may have wondered if I had changed my mind altogether.

Pioneer Ranch looked far more welcoming under summer sun and green fields than it had on that dark, snowy February day, though the scope of what lay ahead was still clear. Fences leaned, outbuildings needed repair and thorough cleaning, and the house—shadowed and timeworn—would soon need a new roof.

The renter was still packing, so we pitched our tents, tended to the animals, and unpacked what we could into the garage. We planned to renovate the house first, knowing it would be easiest while it was empty. Our goal was to make it comfortable enough for the coming winter, and in those early days, sheer determination carried us through the challenges.

Richard and I had bought a Sioux-style tipi from Nomadics, the same company that made the tipis used in the movie *Dances with Wolves*. We planned to live in the tipi while we renovated the house, and later use it as part of our accommodation offerings when we opened part of the house as a bed and breakfast.

Tipi and tents erected, vehicles mostly unpacked, we settled in for our first night of camping. Jalebi refused to leave the Honda. Certain the cat was desperate to pee, I left the door open. As soon as he could slip out unobserved, he disappeared. We would not see him again for months.

We had arrived.

Our belongings were stacked in boxes in the large, detached garage. The animals were fed, watered, and bedded down for the night. So were the people. Before turning in, I took one last look around. The sky was filled with stars. I could see the Milky Way. With no outside lights to interfere, the buildings were dark silhouettes against a vast, starry sky.

That last view before I slipped into the tipi was my first lesson in country dark. In the city, I had known only the half-dark of urban nights, except for occasional camping trips far from artificial lights. That first night at Pioneer Ranch, I gazed in awe at star-dark. In the months ahead, I would learn cloud-dark, snow-dark, rain-dark, moon-dark, and—most magical of all—Northern Lights dark.

The star-filled sky was a comfort on that first night, as I settled into a life I had never imagined for myself.

Chapter Forty-Three

HAVE YOU MET THE WEIRD NEIGHBOURS?

News of our tent city spread like wildfire—from Pioneer Ranch to the village of Horsefly, down Horsefly Road to 150 Mile House, and on into the small town of Williams Lake. The new people from urban Vancouver Island were beyond weird. This was cattle country. They had sheep, pigs, and poultry. Ranching folk drove pickups. The new people had a 1967 Ford truck with a wooden box. Instead of living in a house like normal people, they had set up a tipi and a whole raft of tents. Young people were gathering there. Everyone knows young strangers are suspect. (The young people were part of an international volunteer program.)

The house needed updating. So did the fences and outbuildings. The whole place needed attention, yet these new people were ripping the guts out of the house. Their sheep wandered the yards. Who on earth were these strangers?

I kept thinking of something a storytelling student once said when I was living in Seattle. After some years in a small Rhode Island town, she moved back to Seattle and prefaced her story with, "A small town is where people know more about you than you do, because you only know the truth." Everything we did seemed designed to confirm our outsider status.

Then Fred Eaglesmith's agent called. When Fred had given the first loft concert in our barn at Auchinachie Farm, we spent the preceding week building stairs, clearing out mouse and owl debris, shoring up the floor, building a stage, wiring for sound and lights, and advertising the event. At least a dozen people appeared with hammers, tool belts, and goodwill. Fred was the perfect performer to launch our barn concerts. His music was about rural displacement and the heartache of loss. His ballads went straight to the heart. That first concert was symbolically rich for a couple of transplanted city folk trying to make a go of country life.

This time, though, we had no community to call on and no building we could quickly turn into a performance venue. We worked hours every day ripping the interior walls off the house, determined to move in before snowfall. We told Fred's agent we did not have a bed in the house, nor a building suitable for a concert.

Fred would be on tour, his agent said. He would be playing Prince George and Vancouver and wanted a place to stop in between. This time it would be just Fred and Willie P. Bennett—the finest harmonica player I had ever heard. They would do a no-guarantee yard concert and sleep in a tent.

Adding concert marketing to house renovating, fence mending, animal tending, and cooking for a crew seemed mad, but we were eager to see Fred again. We gulped, agreed, contacted the papers, and posted flyers. On concert night, a couple dozen people showed up. They sat on the lawn behind the house in all the chairs we could gather. Most were curious neighbours, glad of an excuse to check us out. They were cautious at first, until Fred loosened them up with a performance as good as anything they would have heard in a high-priced hall.

They relaxed when he pointed out the obvious. We were odd folk in a place where deep roots and tradition were valued. He gestured to our Jacob sheep, with their brown-and-white spots and multiple horns (they can have two, four, or six). "You see the horns on those sheep?" Fred asked the audience. Then he slowly nodded. "Antennae. These people are Communists, and those sheep are transmitting messages." When the audience stopped laughing, they looked at us less suspiciously.

As the last curious neighbours headed up the long driveway, we lit the dry wood in the fire pit and sat around talking until the day's work settled into our bones and sent us off to bed.

I had been feeling out of place and unsettled. Fred and Willie P. reminded me I was relocated, not lost, and that, in time, I might find home again. One last goodnight to those around the fire, one last look at the starry sky, and I slipped into the tipi to gather strength for the coming day.

I needed the rest. Before the last person emerged from his tent, I would be in Cathryn's Cariboo Kitchen, with the coffee on and breakfast cooking on a two-burner camp stove.

Chapter Forty-Four
CATHRYN'S CARIBOO KITCHEN

We were a curiosity to the neighbours, and a curiosity to our friends. Some of the latter made the trek into the wild to see our new digs. Richard's son, his brother, and a passing parade of friends and family joined us to lend a hand and witness for themselves what had drawn us so far from our city and island homes.

Volunteers arrived, too. We had been part of the WWOOF program on Vancouver Island. Those young "Willing Workers on Organic Farms" had given us the gift of their labour and youthful energy in exchange for room and board. Though we had warned WWOOF Canada that any volunteers would be living in tents, they still sent eager travellers our way. We had barely settled into our tipi when the first arrived.

An immediate challenge was feeding them. We had begun ripping out the kitchen shortly after arrival, leaving me with a two-burner Coleman stove under a blue tarp—my all-weather, outdoor cafeteria. One-pot meals became the specialty of Cathryn's Cariboo Kitchen. Add a loaf of crusty bread and a big green salad, and we could feed a crowd. I cooked with whatever ingredients were on hand. No missing spice, vegetable, or cheese was worth the 90-kilometre round trip to the nearest supermarket. Many of our young volunteers were vegetarians, so I added a second pot to the camp stove and kept everyone happy.

When it was time to peel potatoes, toss a salad, knead bread, or assemble sandwiches for lunch, I could count on helping hands and good conversation. Along with Richard's family and the volunteers, three of my storytelling friends came—Jenni Woodroffe from Australia, Michael Parent and Cynthia Orr from the States. My brother and his wife drove up from Napa, California. Their visits eased my isolation that first summer, far from the urban life I still missed and the island I had come to love. When they asked how I was faring in a place that seemed impossibly remote to them—and to me—I straightened my shoulders, plastered a semi-convincing smile on my face, and rattled on about our plans. When they waved a last goodbye, I felt bereft, but with so much to do, I had little time to brood.

Renovations and farming were materials-intensive. The nearest large hardware store was in Williams Lake, so we were grateful that the village of Horsefly had a well-stocked store only twenty-two kilometres from Pioneer Ranch.

Our first visit there was a happy revelation. Richard was canny with tools and mechanical mysteries. We wandered the small, well-stocked store, looking for the bits and pieces he needed for some plumbing task. The owner asked if she could help. When he explained what he was looking for, she picked up one piece after another, deftly fitted them together, and quickly handed him the solution. Score one for a strong, smart woman. We became grateful regular customers.

A more immediate concern than any repairs or renovations, however, was predators. They smelled our chickens and sheep and figured we had opened a restaurant.

Chapter Forty-Five
THE PIONEER RANCH RESTAURANT

If a fox could wink, the one who turned his head to watch how fast we were chasing him gave us a sly one. Even with a Muscovy duck in his mouth, he easily outran us.

He had already picked off several hens. The free-range chickens that had provided a steady income on our island farm were not an option in the Cariboo, and even our sturdy auction hens proved no match for predators. We were living among wild creatures now, and they found our choice of livestock exceptionally tasty.

While birds disappeared into fox maws, some of our beautiful Jacob sheep fell to a different threat. Overnight, a toxin struck them down. We called the local agriculture office, who suggested two likely culprits: a pond tainted with toxic blue-green algae, or antifreeze left where animals could access it—a hazard I had never considered until that morning. With its ethylene glycol, antifreeze tastes sweet to animals (and small children), and death can come swiftly, or slowly as the kidneys fail. Whatever had killed those spotted sheep acted quickly.

The surviving sheep were on edge. As night fell, coyotes began their eerie chorus. Ravens watched for newborn lambs, relishing the delicacy of young tongues and eyes. Bears, cougars, and wolves roamed the hills, occasionally satisfying their hunger with our pasture-raised animals.

Richard bought bullets. He, his son, and his brother began patrolling. One night, his brother settled into the hay of the barn loft, determined to end coyote predation. He might as well have slept comfortably through the night. The predators waited until he gave in to weariness and then resumed their victory chorus.

Our dream of profiting from our sheep was being eaten almost nightly. I loved seeing wildlife so close, but I hated what they were doing to our animals—and I lost sleep over our role as predators in this wild neighbourhood. Something had to change, and it had to change quickly. Richard proposed buying a guardian dog. Fortunately, an Akbash breeder lived nearby. Bruce responded immediately, arriving the same day with the solution: Jacob, a blond giant of total calm, Bruce's stud dog.

We had purchased Mick, a trained Border Collie, to herd the sheep out to pasture and round them up each night. With his help, we gathered the flock in a pasture close to the barn. Bruce drove his pickup across the bridge and through the gate. Smelling an unfamiliar dog, the sheep went on high alert, clustering together and wary of another predator.

Bruce dropped the tailgate. Jacob jumped out and lay down, quietly sizing up the situation. He inched toward the sheep, pausing whenever they drew back in fear. Crossing the short stretch from truck to flock was slow and deliberate. When he finally reached them, the sheep grazed peacefully around him.

Predation at Pioneer Ranch stopped. We bought a five-month-old Akbash pup, and Jacob stayed on to train him. Three months later, Esau was big and canny enough to handle the job on his own. He lived full time with the sheep. They trusted him completely. By the time he reached his full seventy kilos, he could even lick a newborn lamb without being butted by its mother.

We could tick one challenge off the list—but the list grew daily, like the night a fire broke out in our back woods.

Chapter Forty-Six
FIRE IN THE WOODS

Pioneer Ranch lay on one of the historic grassy openings that had once drawn cattle drovers to the Cariboo goldfields. Our quarter section was mostly pasture, ringed by gentle hills. The timbered southeast corner was our private forest, a magical woodland where moose, deer, bears, and even the neighbours' wandering cattle regularly trampled the fences. After nights spent watching over sheep from prowling coyotes or foxes, the woods already felt like a place of both promise and danger.

One late-fall night during our first year, we were watching a movie in the cabin where Richard's son lived. We had renovated his space first—adding a bedroom, insulation, new windows, wiring, and a roof—turning an old shed into rustic accommodation. It was the only place on the ranch with a working television.

As Richard and I started back to the house, a flicker in the southeast woods stopped me cold. Flames? We both knew how quickly fire can spread out of control. Our small forest—precious for its beauty, timber, and quiet charm—was at risk.

Richard grabbed shovels and chainsaws; his son jumped on the tractor, and they roared across the field toward the glow. I ran inside, dialing everyone we knew within a ten-mile radius. It was after eleven, but there was no time to hesitate. If we could muster enough hands, maybe we could contain the fire.

Help arrived quickly. Neighbours appeared with shovels and determination. And then the truth emerged: the "fire" was a huge pile of scrap wood, intentionally burned by a free-spirited neighbour, ignoring the fire ban but completely confident he could control the flames.

No one scolded us for the false alarm. Instead, they praised our caution. That night, the incident became a gentle joke in local lore —a story of the "townies" who didn't yet know the people who lived around them.

We learned something vital: when you live beyond the reach of the nearest fire department, your best insurance is good neighbours. That night, we knew we had them.

Chapter Forty-Seven
SEARCHING FOR PIGS

As the adrenaline from the forest fire faded, I realized that life at Pioneer Ranch was never going to be quiet. Predators, weather, isolation, and unexpected crises would test us constantly. Yet a rhythm was emerging—a rhythm of long days, hands-on work, and the satisfaction of seeing the results. Each challenge, from foxes to flames, reinforced one truth: in the Cariboo, survival—and enjoyment—depended on persistence, ingenuity, and the patience to find joy in small victories.

Renovations, new livestock, fencing, haying, and cooking for a crowd of friends and family took all our time that first summer at Pioneer Ranch. We had little time for socializing with anyone not living or staying at the ranch. But an invitation to a friend's 50th birthday promised a rare escape—a long drive up one of Cariboo's scenic backroads and a chance to meet new people. Don Gesinger, the architect who had drawn up our renovation plans, was celebrating, and we were eager to join.

First, though, we had to tackle a more immediate task: picking up weaner pigs from K & B Piggery in Lac La Hache. There was no phone at the piggery, so we were told to stop at the gas station for a map. The first attendant knew nothing about any maps, but a helpful customer gave us directions, or so we thought. Three hours of wandering lost through backroads and bush later, we returned to the highway and called the gas station again. This time, the owner was there and said, "Yes, of course, we have a map," but gave us directions over the phone. Her instructions were more puzzling than the first, so we drove back to the station, picked up the map, and discovered both sets of directions had left out crucial details.

Armed with the map, we found the turnoff, but the sign read Triple D Ranch, not K & B Piggery. The drive was 2.6 rugged kilometres over a road that, a month earlier, would have been a mud bog. We jounced along, determined not to turn back without the pigs, but wondering if Old Blue would fall apart on us. When we reached the piggery, we found a tiny trailer beside a large hole. The family had been burned out in March and was starting to rebuild. We did not have the heart to complain about the lost hours, given the monumental task they were facing. Grateful for their resilience, we loaded five piglets and headed home.

By the time we returned to Pioneer Ranch, we were already hours late for the party—and found unexpected company at the ranch. A couple who had once been part of a commune stopped by to visit, sharing memories of the farm and wandering through the overgrown remnants of their former market garden. We spent a couple of hours with them, listening and laughing, before finally setting out for the celebration.

Chapter Forty-Eight
CELEBRATING CARIBOO

By the time we arrived at Don Gesinger's birthday celebration, we were a good six hours late, yet the party was still in full swing. Many guests had likely left, but those who remained lingered, enjoying the company and the warm evening.

The crowd was a mix of artists, professionals, and longtime Cariboo residents. Some had moved to Canada as war resisters, refusing the draft for Vietnam. Others were German immigrants. Most had come from elsewhere in Canada, and a few were descendants of families who had arrived during the Cariboo gold rush and never left.

I drifted through the group, delighted by the stories and the sense of history woven into the place. At one point, I heard a name that made my heart skip: LeBourdais. Even with my limited knowledge of Cariboo history, I recognized it. Louis LeBourdais had been an amateur historian in the 1920s and 1930s, gathering photographs, letters, journals, and memorabilia. Richard, who had long written about the Cariboo gold rush, had introduced me to his work. Meeting someone connected to that legacy felt like stepping into the history we were only beginning to inhabit.

Eager to make the connection, I approached a woman with that last name. "Your roots must run deep in Cariboo," I said, nearly gushing.

She smiled politely. "Oh, no, dear. I've only been here fifty years."

Fifty years, I realized, was still not enough to erase newcomer status. I laughed softly to myself, aware that, as a middle-aged city woman transplanted into rural British Columbia, I had my own hurdles to overcome. Time alone wouldn't guarantee acceptance, and history, however rich, was only part of the equation.

The drive home after the party was quiet, contemplative—a chance to unwind after the day's adventures. And then, suddenly, it happened. A cougar, magnificent and nearly three metres from nose to tail, appeared on the road ahead, moving with effortless grace. Our headlights caught the sheen of its tawny coat. In all our years in the Cariboo, it would be the only cougar we ever saw. The animal vanished as quickly as it appeared, leaving us stunned, awed, and deeply aware of the wildness surrounding Pioneer Ranch.

The encounter reminded me, in a way the party had not, that life here was unpredictable, exhilarating, and sometimes breathtaking. It was a vivid reminder that the land and its creatures had their own rhythms, their own rules, and that being a part of this world required both humility and attentiveness.

By the time we reached home, the stars were brilliant overhead. The piglets were settled, the house quiet, and I felt a mix of exhaustion and exhilaration. Between lost maps, late arrivals, and fleeting glimpses of raw wilderness, Cariboo had already shown us that life here demanded attention, humility, and an openness to wonder. It was a lesson I would carry with me throughout our years in the heart of the province.

Chapter Forty-Nine
CEMENTING MY OUTSIDER STATUS

Many of our Vancouver Island friends were taken aback by our decision to move to an area they thought far too remote. Our friend Marnie Duff had the perfect antidote for my uncertainties: a book by an urban academic and writer who fell in love with an old-family rancher. She moved with him to his ranch on the grasslands of southern Saskatchewan.

Ranching did not come naturally to Sharon Butala. She was the proverbial fish out of water, baffled by the social mores of her new circle, overwhelmed by sky and space and isolation, uncertain how to live without the urban trappings that had given her life meaning.

Oh, how I related. I had moved from Seattle—the city I had sworn I would never leave—to Vancouver Island. New partner, new country, new life. As if those transitions weren't enough, we had bought a small farm and plunged into a life with a learning curve steeper than anything I had experienced.

Just as I was settling in, surrounded by friends who were stimulating company, we bought a small ranch—sixteen times larger than our farm—in an area so rural the nearest village had a few hundred souls and the nearest town topped out at around 10,000. My definition of "remote" underwent a major revision over the next nine years. By the time I left Cariboo, I had redefined "remote" to mean living well away from paved roads, without telephones, electricity, or running water. We had all of those. Even the Internet reached us around 1998. It was hopelessly slow, but still a game changer. In 1994, when our adventure began, I considered anything an hour from a major urban area remote. At Pioneer Ranch, we were a seven-hour drive from Vancouver—assuming no blizzards or icy roads slowed us down.

Sharon Butala's *Perfection Of The Morning: An Apprenticeship in Nature* became my companion. I devoured it in one sitting before leaving the island. In Cariboo, I kept it close during my steep learning curve on the 164 acres of our new home. The book and its author were my first new friends as I struggled with isolation.

When a neighbour invited me to join her book club, I was thrilled. At last, I could discuss ideas, share good books, and perhaps ease the loneliness that shadowed my days.

Since we all lived on rural properties with limited access to new books, the club operated differently from most. Instead of selecting a single book for everyone, each woman brought one to share. We introduced the book, anyone who had read it could chime in, and afterward, we swapped titles for the next month.

I was eager to share Sharon Butala's book. Coming to grips with life as an urbanite turned ranch wife, she had dived deeply into her spirituality and found her rhythm as a writer. I felt her misgivings and longings in my marrow. I found solace in her openness and hope in her gradual acceptance of a new life. My heart beat to the rhythms of the book. It spoke my truth.

I was sure other women on Horsefly Road felt some of the same yearnings and confusion I did. I presented the book with absolute assurance that it would open doors between me and my new acquaintances.

Silence followed my enthusiastic review. Others in the group had read *The Perfection of the Morning* and found it far from perfect. They described Butala as spoiled and shallow.

I might as well have worn a sign that said, "Spoiled. Shallow. Newbie. Clueless." I longed for a hole to swallow me. I had made a major gaffe. These women loved their country life. I was the only one dragging my feet, harbouring stereotypes about rural areas and the people who lived in them. I was not just a newcomer. I was a city snob.

It was not an auspicious beginning for my membership, but the women were kind. In time, the demands of our rural property and the contract work I took on left me little time to participate. By then, I had come to appreciate and admire them. They were exactly the people I had hoped to meet—friendly, welcoming, bright, and knowledgeable. Patiently, they challenged my threadbare clichés about rural life.

In that first meeting, I knew nothing of checking on a pregnant ewe on a night so cold frost instantly clogged my nostrils. I had never stood in silent awe watching the aurora borealis dance across the sky. No mild, friendly cow, bending to lick her first calf, had turned into a fierce mother and flattened me for coming too close. Money worries had never driven me to apply for a job for which I was unqualified, only to discover the determination to succeed within me. So much lay in wait for me to experience.

Years later, a local writers' group, all gifted wordsmiths, invited Sharon Butala to give a workshop in Williams Lake. I was one of the first to sign up. I relished the day. Butala's exercises made words flow like a river, drawing the best out of us.

But it was lunch that branded the day into memory. Heidi Redl, the neighbour who had invited me to join the book club years earlier, organized the meal. She invited me to join her and Sharon. I told Sharon of my faux pas in presenting her book to the reading group. She understood. So did Heidi, whose love of writing was sometimes viewed askance by the ranching community. We laughed, ruefully but straight from the heart. Ours was the laughter of survivors, whose bumpy paths were more interesting and satisfying than anything we could have imagined before life threw curves our way.

Richard and I were always a little out of step with local norms—the city folk with strange sheep instead of cattle, mules instead of horses, and red pigs alongside the pink. Still, our persistence and willingness to muck in earned us a measure of acceptance. We learned that being new didn't mean we would always be outsiders, but that earning our place required patience, humility, and a willingness to be surprised by the land and its creatures.

It was this acceptance—and the reputation we were building along Horsefly Road—that prepared us for the next adventure: the arrival of three unexpected guests, a trio of pot-bellied pigs who would quickly become part of the Pioneer Ranch family.

Chapter Fifty
THE HEIGHT OF OPTIMISM

Having established our credentials as the odd folk on Horsefly Road—the ones with strange sheep instead of cattle, mules instead of horses, and red pigs (Tamworth, an old breed) alongside the pink—we were not surprised when someone arrived with three pot-bellied pigs.

The little porkers had reached full size and were no longer cuddly pets. Naturally, their owners thought of us. Had they called, and had I answered the phone, they would have had to search for another solution. But when three friendly pigs poked their snouts into our hands, they became instant family.

We made them a nest of hay beneath the cabin, set out food and water, and gave them free run of the place. Richard's son, living in the cabin, accepted them remarkably well. Until relatives arrived from Vancouver and slept in the cabin, we had no idea the small pigs snored like trucks grinding gears on a hill.

We named the boar Horace and his two little sow pals, Violet and Jelly Belly. Except when Horace was feeling amorous, they hung out peacefully together. Our menagerie was accustomed to wandering the acreage in the company of a wide range of domestic animals. None of the birds or beasts was the least ruffled by Violet and Jelly Belly.

Horace, however, had his own agenda. Violet and Jelly Belly were never enough for him. The Tamworths were many times his size, but that did not deter our little Don Juan. He would pester whichever sow lay down until she moved away in exasperation. His optimism was astonishing.

Horace's enthusiasm sometimes had unintended consequences. One morning, we found him attempting his amorous advances on a Romanov ewe. She was having none of it, stamping her hooves and letting out a sharp bleat of protest. Undeterred, Horace backed off only to circle around and try again. It was a slow-motion comedy of errors, the little boar's persistence matched only by the ewe's determination to maintain her dignity.

Months later, a pickup drove into our yard. A man climbed out and asked if we were missing a small pig. We looked in the back and saw Jelly Belly—who had somehow wandered more than a kilometre from the barnyard, dragging her ample belly through deep grass, under fences, and across a stretch of bush. She had survived the journey unscathed, avoiding the predators that patrolled the area. That adventure satisfied her wanderlust; she never went walkabout again.

Perhaps Horace was inspired by Jelly Belly's escapade. When he tired of unsuccessful mating forays, he set his sights on possibilities farther afield. One day, we got a call from our neighbours, Bud and Sandy. Sandy had found Horace in the yard, about to come out on the losing end of a battle with her dog. She grabbed the dog and chased Horace away. When she called to say Horace was heading home, she warned us that the randy little pig was hurt.

He did not make it on his own. We searched every inch of the fields and bush between our place and theirs before we found our wayward pig, badly mauled, bleeding profusely, and completely discouraged. We packed him home, doctored his wounds, and kept watch. We thought the brave little heart would pull through. He seemed indomitable. But sometime in the night, between our trips to the barn to check on him, Horace breathed his last.

We wept at the loss and soon stepped away from pot-bellied pig parenting entirely. We never forgot their friendliness, their intrepid spirits, and, in Horace's case, his unrelenting libido.

Life at Pioneer Ranch had a way of escalating quickly from charming mischief to full-on, frostbitten drama. Just months later, we found ourselves testing our nerves, stamina, and luck on a dark, lonely mountain pass in the dead of winter, with cows and sheep rattling around the back of Old Blue—as if the universe were reminding us that country living was never, ever boring.

Chapter Fifty-One
TRUCK DRIVER TO THE RESCUE

The truck broke down on a dark, lonely mountain pass. It was deep winter in 1995, and the air was cold enough to freeze any exposed skin. In the back of Old Blue, our 1967 Ford one-ton, rode two Belted Galloway cows and four Shetland sheep. We were driving non-stop to transport the animals from their home in Fort Qu'Appelle, Saskatchewan, to ours near Williams Lake, British Columbia.

All had gone smoothly until we neared the summit of a pass on the Yellowhead Highway, west of Jasper. At 10 p.m., two tires blew. We were an hour and a half from the nearest garage in Valemount, and the temperature had plunged to -20° C. We turned on the emergency flashers and braced ourselves to flag down help.

The first vehicle to approach was a Yanke truck, whose driver pulled over immediately. My husband stayed with the animals while the driver took me into Valemount. He pulled into the town's only all-night coffee shop and began making calls.

The town's only mechanic was out on a job. Undeterred, the Yanke driver located him and arranged for the mechanic to pick me up. I assured him I could wait at the coffee shop, but he refused to leave until he was certain I was taken care of. Hours passed. We chatted, drank coffee, and stayed warm, but I couldn't help worrying about my husband and the livestock, waiting alone on the snowy pass.

The driver kept checking on the truck. I assumed he was updating a dispatcher, but he was actually alerting other trucks travelling the Yellowhead Highway to keep an eye out for a 1967 Ford and its cold driver. My husband was mystified by the trucks flashing their lights as they passed, later learning they were signaling he was not alone.

It was nearly 1:30 a.m. when the mechanic finally arrived. He mounted loaner tires and drove me back to Old Blue. We completed the drive into Valemount, fed and watered the animals, and found a motel still open. After a few hours of sleep, we bought new tires and resumed the journey home.

The animals came through the ordeal unscathed. So did we—thanks to the kindness of a young Yanke truck driver who turned a frightening night into a reminder that strangers can be remarkably compassionate.

Chapter Fifty-Two
AND THE CAT CAME BACK

When Jalebi disappeared on our arrival in the Cariboo, we feared we would never see him again. A suburban cat would have been easy prey for one of our many predators. Two months later, one of our German volunteers spotted him walking along a fence. A few weeks after that, Richard's brother saw him strolling across the garden.

By early December, snow lay deep. One bitterly cold day, a plaintive meow drew our attention. It came from the crawl space beneath the house. We had stuffed insulation into holes leading to the unheated area, and beside one, a pile of pink, fibrous material revealed Jalebi's entrance.

He ignored my calls at first, but the shaking of a box of cat food brought him running. Though he had always been skittish around people, he ran straight into my arms. He struggled halfheartedly, then settled.

Six months of survival in the wild had made him a mighty hunter. His fur gleamed, though he was a little thin. Snow had driven his prey underground, and he had sought shelter. Though he did not want to be a house cat, the Cariboo winter was too harsh for him to survive on his own. With hay and a blanket, I made a nest for him in the barn.

Jalebi could still hunt rodents that ventured too close, but he now had reliable food and water. My daily visits became a highlight of his day. Gradually, he became affectionate. Instead of rushing for his meals, he would rub against me, purring loudly as I petted him. If anyone else entered the barn, he vanished into his nest beneath the floor logs. From time to time, he disappeared for days or even weeks, but he always returned—plump, sleek, and triumphant.

One winter, the cold proved too harsh for our adventurous cat. I brought him into the house. He arrived reluctantly and refused to let Phantom or Mickey—our two Border Collies—anywhere near him. With snow falling and wind whistling outside, he settled into the domesticated life he had never sought. He did not warm to other creatures in the house, human or otherwise, and mostly stayed beneath our bed. As soon as I came to bed, he joined me, retreating whenever Richard appeared.

When the snow began to melt, Jalebi was happy to return to the barn. Yet the years of adventure had left their mark. Despite regular feeding and care, he grew thin. When neighbours brought us two new barn cats, Jalebi retired from mousing entirely, content to watch the world from his nest. In his final days, he seemed to carry both the pride of a survivor and the calm of one who had known freedom.

Not long after, he died peacefully, having lived a life that was entirely his own—independent, untamed, and fiercely brave. And though he was gone, I could still picture him moving through snow and shadow, a creature of the wild who had always returned home.

Chapter Fifty-Three
COLD COMFORT

Snow was flying by the time we retreated into the partially finished house. It would fall all winter while we worked to keep animals fed, water flowing, fences intact, and the old oil furnace running. We had done our best to insulate the space between the outer siding and the inner logs and had re-chinked every inch of the walls. Still, when a freezing wind blew, it found every crack. Ice formed on the inner walls and windows. That January, our heating bill soared to $400.

We never knew exactly how cold it got that winter. Our thermometer had an outdoor sensor marked down to −60°C, until one night it broke. We prepared ourselves for the cold and were generally comfortable under layers of wool and flannel. Below −25°, any exposed skin turned white on the short walk between the house and the barn.

Frost had become a nightly visitor by early October. That was the signal to move out of the tipi and into the house. We had worked nonstop for months. Our architect friend, Don Gesinger, had designed a spacious kitchen; another friend, Ian Buker, had crafted it from local woods. From the first meal I cooked there to the day I left the ranch, I never tired of that room. It had a country warmth with modern conveniences and the soul of a handmade space. We kept the old wood-burning range—not for cooking, but for warming winter-born lambs. With a towel-lined basket on the open oven door, we could give each tiny life an extra boost.

When we began moving boxes from storage in the garage into the house, we discovered that creatures had already begun the unboxing. Pack rats had raided our belongings—chewing through leather, shredding papers and fabrics, and scent-marking their work. My old sense of identity was in those boxes: the books, clothes, and memorabilia that connected me to the life I had left behind.

Looking back, I see that loss differently. Those possessions had clung like limpets to my soul. Their destruction made it clear my life had entered a new chapter. It freed me to attend to the serious task of reinvention—and freed the garage to become a performance space where we later hosted Canadian musicians such as Gary Fjell-gaard and Valdy.

Renovations had given us a home, though it would always be a work in progress. They also made a major dent in my savings. I began searching for work in town. Jobs were scarce in a community of 10,000. My experience as a French teacher, storyteller, and arts organizer didn't match the few openings available. It was time to find a new direction.

Richard spotted an advertisement for a three-month consulting contract with the Cariboo Economic Action Forum. They needed someone to organize focus groups around the region before developing a Made-in-Cariboo logo. I had never run a focus group nor designed a survey, but the contact person was someone Richard knew—apparently enough to overlook my lack of experience.

Soon I was driving to Williams Lake, Quesnel, and 100 Mile House, talking with ranchers, artists, and small business owners about what made the Cariboo unique. By the end of the contract, we had an attractive logo and plans to promote it locally.

Three months later, I was offered the position of coordinator for the Forum. My knowledge of economic development was even thinner than my grasp of focus groups, but I plunged in. I learned to talk with politicians, mill managers, bureaucrats, and First Nations leaders. My left-leaning idealism got a quick education in diplomacy and compromise. The people I met chipped away at my assumptions about resource workers and business owners. The more I listened, the more at home I felt in this region I'd barely heard of before moving there.

For the next three years, I divided my time between farming and consulting. That led to new contracts and a steadier income. Each project added to the toolkit I carried into the next one. Eventually, Richard and I launched our own consulting business—Grass-Roots—built on the relationships and trust we had earned.

Until then, my world had been education, storytelling, and the arts. I learned the truth of something I had always half-suspected: skills are transferable, especially storytelling. Whether crafting a mission statement or guiding a group through change, every organization was really just trying to tell a new story.

From fencing and lambing to facilitating meetings, my years as a reluctant farmer taught me that "I don't know how" was not a reason to turn down a challenge. I might not know yet, but I could always learn.

All that lay ahead, in the years to come. But that first winter tested more than our endurance. Cold crept into every corner of the house and, at times, into my spirit. I tried to stay busy, but worry had a way of finding the cracks.

One night, when the wind howled through the logs and the stars burned cold above the fields, I went out to check the animals —and to steady myself.

Chapter Fifty-Four
RESCUER RAM

Isolation and money worries were biting me as sharply as the blizzards outside. The house still smelled of sawdust and fresh-cut wood, the furnace wheezed, and every breath seemed to hang in the air like smoke. I kept reminding myself that we had chosen this life, but there were nights when choice felt like exile.

Snow lay deep and unbroken across the fields, a white silence that seemed to stretch without end. The cold had a sharp edge, the kind that drives blood from the skin and thought from the mind. I felt far from myself, living a life that didn't yet fit, a story I could not quite recognize as my own.

I slipped out of the house under the pretense of one last check on the animals. From my Idaho childhood, I remembered how winter's stillness could quiet the world even as it froze the heart. I needed that quiet now—to breathe, to think, to feel something steady beneath the uncertainty.

In the dark barnyard, I sank into a pile of hay and let the tears come. The sky was thick with stars, cold and brilliant as glass. Around me, sheep and cattle lay like small mounds of breath and warmth, unconcerned by my presence.

All but one—Black Boy, the Jacob-cross ram. He moved toward me, slow and sure, and lowered his great head until it rested against my chest. His weight was steady, his breath warm. Gradually, my sobs eased. Despair thinned, replaced by something quieter—something that felt like peace. He seemed to draw the ache out of me, shaking it off from horns to hooves.

Warmth returned, first to my chest, then to my spirit. I reached up and scratched the places he loved—the woolless spot where his legs joined his chest, the soft curve behind his ears. I don't know how long we stayed that way, only that time loosened its grip. The stars kept their vigil, and the cold seemed less cruel. When I finally rose, I could move forward again.

After that night, Black Boy always found me. If I was content, he would stand close, patiently waiting for a scratch. If sadness weighed on me, he would linger until I sat, then press his head against my heart as if to remind me where I lived.

In my years in the Cariboo, many animals made me laugh, tested my patience, and softened my heart. But only one reached me when I was lost in my own silence and showed me that even in the coldest season, life still held warmth.

Chapter Fifty-Five
THE GRIEF OF A GREY GOOSE

When we moved from the island to Cariboo country, we brought the geese with us. By then, friends had given us a third Toulouse. They were amiable and contented in their new home—until Suli, our Akbash guardian dog, gave birth to her first litter. The growing pups loved to make the geese honk and run. In disgust, the three birds flew into the swamp. Before we could retrieve them, a predator dispatched one.

As the pups matured and gradually accepted our insistence that geese were not for chasing, a truce grew between Lucy (the female goose), her mate, and the young dogs. Akbash are bred for livestock protection, so the geese soon became part of their charge. One by one, other farms bought pups to guard their stock or their families, until only one remained.

One day we found the Toulouse male dead beside the remaining pup. There were no marks on the bird, so we never accused the dog. The goose was not young when we bought him and appeared to have died peacefully of old age. The pup was likely just standing guard, waiting until we came. That left Lucy alone again.

She grieved, as before, then settled into her solitude. Through the long seasons, she seemed content enough—until the sky began to fill with wings. During the spring and autumn migrations, when the great V's of Canada geese passed overhead or rested briefly to feed, Lucy's composure dissolved. Her mournful cries rose to meet the wild calls, echoing across the pond, as if she could will herself into their company.

Her cries pierced me. Lucy's hunger for a life whose call she could not answer felt too much like my own. She was an earth-bound goose yearning for flight, stirred each season by a reminder that rekindled her loneliness. And I, an urbanite trying to be a good farmer and a good wife, felt the ache of a different kind of exile, knowing that only hard choices could return me to a life that truly fit.

So I did the one thing that steadied me, the act that has always helped me find my way through sorrow: I wrote about it.

The Grey Goose

The grey goose lifts her chest
And calls to wild relatives
Who fly over the pond
Where she keeps vigil
Since the morning her mate lay dead
Beside the dog's paw.

Last fall there were three,
Together five years
Until the pups plagued them so.
They waddled and flew into the swamp.
Only two returned,
The grey and her Toulouse mate.

When young she partnered with the African
While the Toulouse played worried uncle

To the goslings.
Wanting purebreds, we sold the African.
The grey cried for a week
Before accepting comfort
From the worried uncle.

Five years he guarded her,
Kept watch as she grazed
Or sat a nest on eggs
That never hatched
But faithful,
As if this year they might.

Twice mated, twice bereaved,
She swims the pond alone
Till summer brings her
Bufflehead and goldeneye.
Hardest are the seasons' shifts,
Flying V's that neither dip nor wait
Nor answer when she calls.

I know times like hers,
Here in my country home,
When the call of other lives, other loves
Makes me feel a stranger
Among boots and saddles.

When I finished the poem, I felt as if I had given voice to both of us, the grey goose and the woman who watched her. Lucy still called to the skies in spring and fall, and I still struggled to answer my own restless longing, but something had softened. Between her earthbound patience and my searching heart, there was a glimmer of understanding.

Chapter Fifty-Six
THE INDOMITABLE RED HEN

Chickens were never a major part of our farming in Cariboo. Predators found our hens and ducks far too tempting. Still, the survivors were a canny lot, outsmarting ravens, hawks, and the occasional fox. Once we brought in guardian dogs, predation stopped, and the birds had little more to worry about than their next meal.

Best of them all was Bonnie, a Rhode Island Red with a fondness for human company—mine in particular. Whenever I came near, she would crouch, waiting for me to scoop her up for a snuggle. Once, she even rescued me from a Silkie rooster who objected to my egg gathering and came at me, claws outstretched. Bonnie squawked a warning, then charged him full tilt. Twice his size, she won the battle easily. The chastened rooster never attacked again.

It was Bonnie, too, who raised two Canada geese from eggs to goslings. In her last winter at Pioneer Ranch, she spent the coldest spells indoors, coexisting peacefully with the dogs and whatever baby animals needed extra warmth.

The rest of the flock led seemingly contented lives, ignoring us unless we arrived with grain or fresh greens. In spring, they gathered around as I turned the garden soil, darting in to snatch up worms and beetles. In summer, the fuzzy white Silkies helped keep the garden free of pests. I missed their company when snow drove them into the coop for warmth.

We had been at the ranch a year and a half when Richard came in from morning chores with startling news: he had heard a chicken under the tool shed. The temperature was well below freezing, and the snow lay deep. I couldn't imagine how a hen had left the coop and found her way there.

When I checked, I found one of our hens—certainly not Bonnie, who had more sense in winter—had gone broody. She had chosen the dark space beneath the shed to lay her eggs and somehow dug her way in.

I snagged her when she emerged and popped her back into the coop, where we provided food and warmth to keep the flock healthy. She pecked at the grain, drank, and escaped to her hidden nest.

The eggs were too far out of reach to retrieve. Their only insulation was the snow piled high around the shed and the hen's warm feathers. After several days of returning her to the coop, only to watch her disappear again, I gave up. I was sure the bitter cold would freeze the eggs, and she would eventually return a chickless—but wiser—hen.

Mick, our Border Collie, always joined me for morning chores. He loved the chickens, their bustle, their chatter, their constant motion. Each morning, when I said "Chickens!", he would race to the coop and wait for me to open the door.

One morning, though, he wasn't at the coop. He was beside the tool shed, eyes fixed on a small hole in the snow leading to the hidden nest. The hen hadn't emerged for her feed, and I feared she had frozen in the night.

I called Mick to come, but he wouldn't move. He had a strong sense of order, and though he had tolerated her independent streak, he could not accept her silence. He knew she was still alive and in trouble. He was insisting I help her out.

He stayed by the hole while I fetched some grain and water. When I gave the word, he dug his way beneath the shed and gently herded the hen toward me. She was thin and bedraggled but clucked with relief when she saw the food and water. Mick stood guard as she ate and drank before disappearing once more beneath the shed. I tucked snow around her entry to hold in what warmth she could make with her small, determined body.

A week later, she emerged on her own and returned to the coop. She had done her best, faithful to a dream that could not hatch. Watching her, I felt a kinship with that small, stubborn heart—both of us clinging to warmth, believing that love alone might be enough to carry us through the cold.

Chapter Fifty-Seven
WILD GOOSE SUMMER

The annual winged migrations used our fields as resting stops on the way to nesting sites. Canada geese were among the regular visitors. We would hear their cacophony for a day or two. At some internal signal, they would rise skyward, form a perfect vee, and fly north, leaving space for the next flock.

Other than a pair of sandhill cranes who stopped regularly to hatch a chick or two, the migrating birds never lingered to raise a family. Perhaps that's why one goose couple who decided to nest in our field met such a violent end. They were a tempting dinner for the many predators patrolling our land: coyotes, bears, wolves, cougars, foxes, and ravens.

On a wander to inspect one of our hayfields, Richard found an abandoned nest with two warm eggs. The geese had fallen victim to one of the local hunters, but their eggs might yet survive. He scooped them up and brought them to the chicken house.

By then, our flock was small thanks to hawks and weasels, but we still had a handful of hens and one rooster. Bonnie, my favourite, was a broody hen who loved attention. She was always eager to be picked up and cuddled.

Nothing pleases a broody hen more than a clutch of eggs to tend. When Richard tucked the two monster eggs into her nest, Bonnie didn't hesitate. She hopped up and settled onto them. Personal comfort be hanged—she had a job to do. She clucked contentedly as she balanced precariously, a tiny sentinel over her unusual charges.

For the next month, unless she took a break for feed or water, Bonnie remained on those eggs. Then, one by one, two little goose-billed, fluffy chicks peeped out from under her wings. Mothering began immediately.

Neither hen nor goslings seemed to care that they were different species. Bonnie taught her foster chicks to eat grass and other delectable bits. They followed her everywhere. At night, they snuggled under her warm wings. Within weeks, they were as large as she was. One day, they wandered near the pond. They scrambled into the water, swimming, while Bonnie squawked warnings from the shore.

When fully feathered, the goslings began making short flights. Bonnie was beside herself each time they lifted into the air. The young female made perfect landings from the start. The male, however, would skid to a somersaulting halt or plow into a fence, honking until we rescued him.

By September, the goslings had grown into magnificent adults, yet chickens remained the only family they knew. We pondered nature versus nurture and worried that our cold Cariboo winters might be too harsh for wild geese.

As the days grew shorter, the first vees flew overhead, heading to warmer climes. Bonnie ignored them, but her two fosterlings gazed skyward, adding their honks to the passing cacophony. Occasionally, a flock landed on the stubble of our hayfields to feed, drink, and rest before continuing their journey.

One morning, our geese were gone. We wondered how they would fare among a flock of strangers. The following spring, one of the migrating flocks stopped in our hayfield. Before settling for the night, two geese walked up to the barn. I fed them a scoop of grain. They gobbled it eagerly, then rejoined the flock.

We never saw them again. Or, if we did, they were so fully integrated into their wild life that we no longer recognized them. We were lucky. Anything could have happened that wild goose summer —predators might have made a quick meal of the goslings, our intervention might have interfered with their destiny, or a conservation officer might have frowned upon keeping wildlife.

Nothing untoward happened. I remained forever grateful for that summer, when we had the honour of living so closely with two wild geese, witnessing their first flights, and glimpsing the delicate, fleeting threads that connect humans to the wild.

Chapter Fifty-Eight
DEATH VISITS THE RANCH

The huge cross atop a cabin's snow-covered roof, outlined in red Christmas lights, confused me. Some houses in Williams Lake sported similar symbols. Cariboo was still new land to me that first Christmas, but I had assumed Canadians all agreed the season celebrated the baby's birth, not his later crucifixion. Perhaps it was a regional custom. I did not remember ever seeing Christmas crosses on Vancouver Island homes.

That was only one of the surprises winter had in store. Our determination to preserve the genetic pool of old breeds of sheep had undergone a harsh reality check. With one small barn and one large pasture, we had no choice but to let our rams have their pick of the ewes. They were still on Island time that first year, which meant lambing in January, when temperatures were cold enough to turn wet newborns into little lambcicles.

Nature is a miracle worker. With only the warmth of hay and the body heat of the ewes, plus the occasional heat lamp on the bitterest nights, most winter-born babies were licked dry and struggling to their mothers' teats before frost could form on their birth-gummy coats. At least, most of them.

One winter, Spats, a Suffolk ewe and one of our most reliable mothers, backed into a corner to give birth. Of the four corners in the lambing pen, she chose the one with the water bucket. By the time we checked on her, a perfect lamb was frozen stiff. She had moved forward to deliver the next two. Fully devoted to the survivors, she wasted no time mourning the lost lamb, attending immediately to those already butting at her udder.

The bitter cold sometimes forced us to bring the weaker newborns into the house, warming their tiny bodies on heating pads or the open door of the old wood range. Once warmed and fed a bit of lamb milk replacer, they were generally strong enough to be in a stall with their mothers until the weather allowed them to thrive outdoors.

We had acquired a few Romanov sheep, known for hardiness and multiple births. One gave birth to twins on a bitter winter's day. We carried them into the barn, but the smallest, black with white markings, struggled to use his legs. We worked with him in the house and barn until he was strong enough to compete with his sibling at the teats. Both soon romped around the stall and occasionally escaped into the passageway. They seemed perfectly healthy and vigorous.

We were pleased with their progress until the day Ramona, a young German woman spending the winter with us, ran into the house sobbing. We followed her to the barn and found a horrific scene. We had placed a sow and her new piglets in the stall beside the ewe and her twins. One lamb had squeezed through the rails and wandered too near the piglets. In a single bite, the sow had decapitated the intruder.

Our gentle young friend had witnessed the carnage but was helpless to intervene.

Guilt came to visit us that day. In our eagerness to give both lambs and piglets a good start during the cold spell, we had failed to consider the risks of placing the ewe and sow in adjoining stalls. The lamb—and our stricken guest—paid the price.

That night, as the wind sifted snow against the windows, I reflected on how thin the veil is between care and harm. Life on the land was a constant weaving of birth and loss, beauty and brutality. I wanted to believe that love and attention could keep the balance tipped toward gentleness, yet the acreage had its own fierce rules. Even so, I resolved to keep trying—to mend what could be mended, to learn from what could not, and to find grace in the uneasy bond between our human hearts and the natural world surrounding us.

Chapter Fifty-Nine
THE DEEP SOUL OF A DOG

Suli came into our lives in 1997, when we decided we wanted a second guardian dog and considered the possibility of breeding Akbash for other livestock owners. She had spent years guarding sheep for good friends on Vancouver Island. While Esau, at seventy kilos, was at the large end of the breed, Suli, at thirty, was on the small side—lean, graceful, completely devoted to her flock, and shy around humans. Born on a Montana sheep farm, she had not seen people in her first four months. A visiting vet told us that was a crucial bonding time, one she had missed.

Still, we knew she had loved our friends, and we hoped we could earn her trust. I took on the task, grateful to have two dogs to watch over our sheep. We had heard stories of coyotes luring a guardian dog away while the rest of the pack attacked the unprotected flock. With two on duty, one would always stand watch.

When I fed the animals, I came as close to Suli as she would allow, never meeting her eyes, always altering my path when she stiffened. I brought treats for both dogs and distracted Esau while Suli cautiously edged toward hers.

Months passed. Suli's comfort zone expanded. I remember the

day I felt her nose brush my hand. I froze, not looking down, while she sniffed my leg, my shoe. Weeks later, I held out my cupped hand with a treat. I looked away as she delicately plucked it and sniffed for more.

It was a turning point. That same day, she allowed me to touch her chin as she took another treat. The next day, I could kneel beside her. Day by day, her reserve melted until I could stroke her neck, scratch her ears, wrap my arms around her.

Then came the day she ran to me, tail wagging, eyes bright, snuggling her body into mine. I wept. My gratitude never waned, even when I had to re-earn her trust after being away for a few days —or worse, a few weeks. She was an exquisite soul: steadfast, pure, wholly devoted to the task for which she was born.

All she needed was sheep to guard. Humans were peripheral. Years later, when we sold the farm, we insisted the new owners keep Suli and enough sheep to keep her happy. They agreed, and in time, they too made friends with this extraordinary dog.

Suli reminded me that love need not tame what it cherishes. Some hearts are meant to roam the edges, guarding, watching, belonging entirely on their own terms.

Chapter Sixty
PHANTOM PAYS THE PRICE

On Auchinachie Farm, our goofy Border Collie, Phantom, led a constrained life. He was a happy wanderer, often racing to the nearby school to find children eager to play. Our fences were inadequate for a roving canine. If he wanted to be outside when we were busy inside, he had to be on a chain.

Guilt was always my middle name when it was my turn to clip him onto the chain. The move to Cariboo and a small ranch meant freedom for him. From June, when we arrived, until early February, he was a gloriously happy pet. He could come and go as he wished. With 164 acres to explore, mice to catch, and humans to love him, he was in doggy heaven.

Every morning he raced to the top of the driveway, sniffing along the way. Then he ran back down to bark a lively report about the creatures who had left their marks overnight.

He was an excellent guard dog. Anyone who drove onto the property was met with a flurry of barking. Although he wagged his tail in a friendly welcome, he was big enough to look dangerous.

Like most loved dogs, he was happiest in the company of his people. He viewed Mick, our smart Border Collie, as a nuisance and rival for our attention. He liked Esau, the Akbash guardian dog, but was always wary after the bigger dog attacked him for sniffing the interesting end of our Akbash female.

Mostly, though, he was just a dog in love with life. Had he been a guy, he would have been a beer-swilling, joke-telling buddy, never accomplishing much but always good company. Pioneer Ranch was his bar. Anyone who came was his best pal. Life could not have been better for him.

Then our neighbour's dog came into heat. Her perfume wafted over the fields, down the hills, and into Phantom's eager nostrils. He was enchanted. He was seduced. He answered the call.

We had let him out for his last pee. It was a frosty winter's night. Normally, he returned within a few minutes. This time, a quarter of an hour passed with no scratching at the back door. We called his name. No answering bark met our ears.

We pulled on warm clothing and went searching, calling his name across the fields. We feared the worst—that he had fallen victim to one of our many predators, that he had become disoriented and lost, or that he had injured himself and might freeze before we could find him. We finally gave up and fell into bed with heavy hearts

At three in the morning, he barked to be let in. He bounded into the house with joy and seemed puzzled when our usual friendly greeting was somewhat tempered.

Our neighbour was less than pleased that our mutt had tampered with his purebred. We were not impressed that Phantom had chosen to mate with a dog reputed to have killed one neighbour's dog and mauled another.

Phantom was oblivious to our worries. He had tasted one of the finer things in a dog's life. We knew that, if he wandered again, he might be shot. Livestock owners are rarely forgiving of roaming dogs, no matter how friendly they are.

So I called the vet and made an appointment to have Phantom snipped. We should have done that earlier, but money for an optional vet bill was always needed more urgently somewhere else. Phantom had moved the expense to the top of the priority list.

So I called the vet and booked the first available appointment. Ironically, it fell on Valentine's Day. We were relieved when the procedure returned Phantom to his contented farm-dog self, whose wanderings ended at the edges of our land.

We were performing cowboy poetry when Phantom was seduced, so I wrote this one about his escapade. Any blame to coyotes was pure poetic licence.

Phantom's First—and Last—Spree

He wanted out at nine o'clock,
To answer nature's call.
Coyote sang a lover's song,
Inviting him to fall
Into a life of wandering
With canines half his size.
This night his rite of passage
Would be through glowing eyes.

He'd never had a girl before.
His life had been sedate.
He'd wandered with his human pals
But always celibate.
The cry that lured him out that night
Just could not be denied.
He ran out barking eagerly,
A new spring in his stride.

Back home his humans fretted.
They feared he'd be attacked
By coyotes seeking sheepdog flesh.
While wandering in a pack.

They paced the floor; they walked the fields.
They trudged up icy hills.
Midnight, at last, they heard his bark,
Made joyous by new thrills.

He wouldn't heed his humans' cries,
Intent on better games.
With heavy hearts they turned back home,
Not knowing if the strains
Of coyote music would win out
And keep him running free,
Perhaps prey to a rancher's gun
As payment for his spree.

At 3 a.m. he sauntered home
And barked to be let in.
He wagged his tail as if to say,
"Have you been worried, friend?"
Today he slept the whole day through,
Not knowing what's in store,
His Valentine's date is with the vet.
He'll have those sprees no more.

Phantom settled back into his days of companionship and play, content at last, no longer tempted to seek adventures beyond the boundaries of the farm.

Chapter Sixty-One
BEST FRIENDS UNTIL THEY ARE ENEMIES

Guardian dogs brought their own set of challenges. As with everything else we undertook, we thoroughly researched the breed (Akbash) and their purpose (guarding livestock, especially sheep). Experts repeatedly cautioned that we must assert ourselves as the dogs' masters. Otherwise, in a situation where they might see us as threats to their flock, they could turn on us. For example, if we needed to treat a sick or injured sheep, the dogs had to respect our position in the flock, even if the sheep cried out.

The experts backed their advice with stories of owners who had been maimed or killed by their guardian dogs. Looking back, I suspect such outcomes are rare when a flock enjoys constant human contact. On our quarter section, we were close to every animal every day.

I took the warnings seriously as our first Akbash, Esau, grew from a rambunctious pup to a headstrong adult. Though I had been hopeless at training Phantom, I learned to be firm, consistent, and, at the same time, loving with Esau.

The truth was, of course, that Akbash need little instruction. The breed comes from western Turkey, with millennia of sheep-guarding experience embedded in its instincts. Esau only had to learn to trust that his human handlers were in charge and would act in the best interests of the flock.

When fully grown, Esau was twice Phantom's size. They both had full run of the farm, but Esau stayed with his sheep. When Phantom came into the barnyard or out into the pastures, the two would romp and play until we returned to the house. Then Phantom would follow us home. Esau never saw the smaller dog as a threat and enjoyed his company.

Then Suli came into our lives and quickly let Phantom know he was no longer welcome among the sheep. He could follow us to the barnyard, but he could not pass through the gate.

Everything went smoothly until Suli's first heat. The three dogs were outside—the Akbash in the barnyard, Phantom near the house. Perhaps Suli grew annoyed at Esau's insistent attention. Whatever the case, she wandered into the yard behind the house. Esau followed her. Phantom, though neutered, likely offered the usual sniff greeting when Suli approached.

All hell broke loose.

We were in the house when the howls and snarls erupted. Richard ran for pepper spray and a rifle. I ran to the dogs. They were so thoroughly entangled that Richard could not have gotten off a clear shot. At half Esau's weight, Phantom was already bleeding and losing ground. Had we not heard the commotion, we might have lost our beloved pet.

Acting purely on instinct, I ran to the two dogs. Grabbing Esau's tail with both hands, I jerked hard. Only afterward did I consider what might have happened. Adrenaline pumping, hackles raised, Esau might have turned on me—or he could have ignored me and continued tearing into Phantom.

The tail pulling startled him. Esau whipped his head my way. Everything I had read about human-guardian dog dynamics clicked into place. He saw me as the one in charge, and the instant recognition caused him to release Phantom. Ignoring the smaller dog he had been attacking, Esau focused on restoring our relationship. Richard scooped up Phantom while I guided Esau back to the barnyard. Together, two humans and one badly bitten dog made an emergency run to the vet in town.

Phantom healed. Esau never turned on him again. Suli had her pups. She had a second litter the following year, and then we had her spayed. Both sets of pups accepted Phantom as just another part of their big, diverse family.

That night's horror never left me, but the dogs moved on quickly, as dogs do. Suli remained devoted to her sheep, keeping Phantom away but never threatening him. Phantom learned to steer clear during her heat, and Esau remained as friendly as ever.

Love and trust held steady at the perilous border where instinct and devotion collide.

We found good homes for the pups, and all seemed well—until the day a vehicle came down our driveway and shattered my peace of mind.

Chapter Sixty-Two

"DO YOU WANT HER BACK, OR SHALL I SHOOT HER?"

Tires crunched on the gravel drive. A red van pulled up and stopped. I stepped outside. The man who got out was the father of the young family who'd bought an Akbash pup from us nine months earlier.

"Do you want her back, or shall I shoot her?"

He opened the back of the van and let out a thin, cowering dog. Her teats were swollen with milk. She hung her head, tail tucked tightly between her legs. The joyous pup we had said goodbye to nine months earlier was now a frightened adult.

The family had made several trips to the ranch, searching for a guardian dog to protect them from predators. We had one who seemed a good candidate—a gentle little female who preferred human company to sheep. We had our reservations but couldn't quite pinpoint them. The children were calm around the pup. The parents seemed reasonable. Finally, we sold the three-month-old female, with the proviso that they have her spayed as soon as she was old enough. They drove off happily, and we hoped we had made the right decision for the young Akbash.

The dog who stepped from the van was a shadow of that joyous pup. Seeing her swollen teats, I spluttered, "You were supposed to spay her."

"She got loose. I drowned her pups in a bucket. She barks. She runs away. She's disobedient, and she bothers the neighbours."

He climbed back in the van and drove off.

I slipped my finger into the dog's choke collar. She followed obediently. Wanting to give her time to adjust, I put her in a separate kennel with Blue, a neutered male and the only unsold sibling. He was a sweet-natured dog, and the two settled in easily.

I began working with her, taking her along when I fed the sheep or checked fences. She never strayed from my side and responded readily to simple commands. She and Blue returned to the stall willingly, but she always looked at me as if she feared I might never come back. When I did, she leapt up and wrapped her legs around me. Her uncut nails carved grooves on my chest and arms.

I wrote to a friend: *She has wrapped herself around my heart. She needs a family who will love her beyond reason and still the insecurity that makes her fear the love will be withdrawn.*

For weeks I didn't name her, afraid of becoming too attached. Finally, a friend suggested I call her Shira—Hebrew for "poetry" or "song." She was making my heart sing, and the name fit perfectly.

I was determined to keep her until I was certain a new owner would love her unconditionally. Diane Dunaway, a friend who volunteered and raised funds for the BCSPCA—and who had one of our pups—kept watch for the right person. Then one day she sent an SPCA volunteer to meet Shira. The young woman spent time with the whole Akbash family and then alone with Shira. She fell in love. So did Shira.

Shira trotted off with her new owner, tail high, ears alert. She stopped once to glance back at me, then hopped into the truck.

I made the woman promise to return her if it didn't work out. "We'll make it work," she said—and she did. After three months of puppy freedom, nine months of terror, and two months of healing,

Shira finally had a family to love her as she deserved. She became a devoted guardian dog and a cherished companion.

I felt lonely for Shira after she left. So did Blue. He kept looking for her in the kennel, then running back to me. When I invited him on a walk, he pulled away and returned to the space they had shared. He settled down in the hay and looked at me with big, sad eyes. That night, he howled his loss.

I mourned her soft presence. But Shira had what every dog deserves—a forever home, and the freedom to trust again.

Chapter Sixty-Three
"I DIDN'T KNOW YOU WERE EDUCATED"

Field by field, we improved the hay we grew and sold. In winter, we harnessed the mules to the stone boat—a wooden platform on sled runners—and hauled hay out to the pastures. A line of cattle and sheep followed, eating greedily and trampling manure into the feed. Come spring, the rotting hay and manure fertilized the fields. The soil, and our hay crop, improved steadily.

Before we switched from forty-pound square bales to machine-handled round bales, most of our hay went to horse owners. One regular couple had government jobs, a late-model truck, and salaries far higher than two freelance farmers could dream of. They would stay in their warm cab while we loaded their truck, and I never guessed they might be quietly judging my practical attire.

One year, I was asked to help judge a spoken-word competition. The local newspaper printed biographies of the judges, detailing our professional and educational accomplishments. A few days later, the female half of the hay-buying couple came out of the bank as I walked by. She greeted me warmly, then exclaimed: "I didn't know you were educated."

The phrase still rings in my ears—not just because it revealed the hay buyer's assumption about the farmers growing her horses' feed, but because it reminded me how I, too, carried my own city biases into farm life. Rural living shattered most of them, though a few lingered. That kind of unlearning takes a lifetime.

A writer has the means to handle slings and arrows. When our hay customer tipped me off to the impact of my work clothes on her perception of me, I turned to words. We were performing on the cowboy poetry circuit at the time, so I wrote this poem. Richard often performed it—switching pronouns as needed—while I dressed piece by piece in my actual work clothes. The country folk in our audiences laughed in instant recognition.

Fashion Statement[1]

Come winter in the Cariboo[2]
I make a fashion statement.
By wearing duds a rancher's wife
Can find in bargain basements.

For every rip and stain contains
A story, maybe two,
About what ranch life's really like
Up here in Cariboo

My warm wool socks were spun
And knitted by a neighbor's hand.
No, not the wife's; the pattern here
Was fashioned by a man.

Next come the ski pants, though on me
They've never slid a hill.
I split the seams while pulling calves
And throwing hogs some swill.

My man says I should toss these pants.
The fabric has gone rotten.
But none of those at Sally Ann
Will fit my queen-sized bottom.

My wool coat's lost the zipper,
And the buttons barely cling.
But it's seen front-line duty
And for chores this coat is king.

The stain right here? It's from
A newborn needing extra care.
His mum had two but didn't want
The second of the pair.

We know we should have left her.
Those cow mamas know the score,
But we can't let a baby die
Just 'cause it's doing poor.

The boots cost me six bits,
And now the sole's begun to split.
But when I don three pairs of socks,
They're still a perfect fit.

I could, of course, buy new ones,
But they cost a lot of dough.
We spend it all to keep
This operation on the go.

On windy days I wear a scarf
To keep my neck from freezing,
Not Isadora Duncan-style
But tucked, unless I'm sneezing.

Then one end is a tissue,
And the other is a rag.
I wipe my nose and dry a calf
And clean the new mum's bag.

My cheeks and nose have been frostbit
So when the weather drops
To minus 30 Fahrenheit
I pull out all the stops.

I wear a balaclava
Like a thief in cheap disguise.
The only things exposed are just
Two nostrils, lips and eyes.

The nostrils drip; the mask
Gets crusted up with salty ice.
I scared the barn cat so
The startled feline scratched me twice.

My mittens have more colour now,
From cleaning newborns' bums.
I toss them in the rag bag
Till the year's first blizzard comes.

It's when I don my hoser[3] hat
My outfit is complete.
I look the part I'm playing
From my head down to my feet.

I wouldn't win a prize for style,
This chore-outfitted wife.
I'd pass for Michelin tire's man
But it suits our way of life.

The laughter that greeted "Fashion Statement" was never mean-spirited. It was the laughter of recognition—the sound of people who knew exactly what it meant to choose warmth over style, to wear the story of their labour on their backs.

On stage, in my patched and mismatched layers, I wasn't a city transplant playing at rural life. I was simply one of them.

Farming had stripped away so many illusions—about dignity, control, even mastery. What remained was a quiet reverence for the lives entrusted to us.

---

1. First published in the May/June 2001 issue of *American Cowboy*.
2. Cariboo is a ranching region in central British Columbia. The first ranches were started in the mid-1800s by Americans, who drove cattle north to feed the gold miners.
3. "Hoser hat" refers to a plaid wool cap. The term "hoser," popularized by Doug and Bob McKenzie's *Great White North* comedy sketches, became a joking term for working-class Canadians—affectionate when self-used, but an insult when applied to others.

Chapter Sixty-Four
BEARING WITNESS IS A GIFT

We were always the outsiders—city folk learning, awkwardly at first, how to live by the rhythms of the land. Over time, we earned a measure of respect, but most of our lessons came not from people but from the animals and the quiet, persistent cycles that governed their lives.

As I had many nights before, I walked to the barn to make one last check on a pregnant ewe. She was an experienced mother. My visit was as much about my own need for reassurance as it was about trying to be a responsible farmer.

That night, things were not going well. Two tiny hooves and a sweet nose were visible. It looked as if the lamb was positioned correctly and ready to whoosh out of the birth canal. But as I watched, the ewe strained unsuccessfully. I could see she was exhausted, so I went into the house to scrub up and ask Richard to come with me in case I could not pull out the lamb.

My hands were smaller, so I was the one who felt around in the birth canal, trying blindly to figure out why the lamb was so stuck. The ewe pushed. I pulled. The lamb remained firmly lodged.

My husband took over. Though his hand and arm were larger, he discovered what I had not: twin lambs hopelessly entangled in the birth canal. Without help, both would die.

Patiently but quickly, he disentangled the two lambs. As soon as their limbs were separated, he tugged on the first set of feet, then the second.

The procedure broke one of the second lamb's legs, but by then Richard was an old hand at splinting animals. Before we left the barn, both lambs were suckling. The one with the broken leg was balancing easily on three. Well before the newborns joined their leaping cousins, the injured one's bone had healed.

It was one of many moments of grace in a life that always felt foreign to me. I, who had embraced city living and never wanted to live in the country, found myself repeatedly playing midwife to a ewe, witnessing the birth of a litter of pigs, or bottle-feeding a weak calf. I will always be grateful for that.

Not every experience with lambs turned out so well. One gentle ram lamb was in distress when I did the morning chores. By nightfall, he was dead. The young rams had been frisky with each other, learning how to butt as if they were already competing for ewes. Rams have thick skulls that spread out the impact. Most of ours had horns, which act as natural shock absorbers. Their neck muscles are strong. They learn to butt each other head-on, but the youngsters sometimes miss their targets. The sweet ram who died may have had his spine broken—not out of malice but out of inexperience on the part of his attacker.

With a sad heart, I walked among the sheep and cattle. Black Boy, the old ram, came running for his scratch, and the Shetland who had taken his place stood quietly nearby. Their calm presence reminded me that life went on, even in the face of loss.

We did what farmers do. We called a friend to help butcher him and prepare him for the freezer. Birth, life, and death—all threaded through our days like the changing seasons, constant and unyielding, yet somehow full of grace.

Chapter Sixty-Five
WILD NEIGHBOURS, WILD LIFE

Living with wildlife was both a blessing and a curse—though, over time, the blessing prevailed. When we moved our sheep and poultry from a suburban farm to a rural property, the local predators were thrilled with the new food source. I could grudgingly understand coyotes snatching a sheep or a fox running off with a hen. But the ravens I had always admired moved to my enemy list when I saw them pecking at the eyes of newborn lambs. Only much later could I forgive them their nature.

The balance gradually shifted for me, from uneasy truce to spirit-lifting appreciation. Spring and fall brought migrating flocks of geese, mallards, and wigeons. They rested, drank, and feasted on whatever seed heads and grass they could find. The next day they flew on, to be replaced by other flocks. A pair of sandhill cranes stayed each summer to raise a chick or two before moving on. Killdeer ran their broken-wing feints to lure us away from their nests. As the years passed, the rhythms of migration and return became my calendar.

One winter, two moose lingered in the pasture just beyond the barn, probably feeding on our hay while we slept. On a bitterly cold night, I saw their dark silhouettes in the trees, motionless and watching. The guardian dogs, knowing instinctively they posed no threat, stood silent. I finished my chores under their gaze, feeling small and strangely blessed in their company.

The guardian dogs gave me freedom—to appreciate the eerie chorus of coyotes, the silent tread of cougars and wolves, the reconnaissance flights of hawks and ravens. Because of them, I no longer felt the old fear of predation—from wild creatures or from the humans with guns who hunted them.

Overhead, winter offered the most magical sky show of all. Sometimes the Light Dancers began their performance before I went to bed. No matter how cold the night, I would pull on warm boots, scarves, gloves, hat, and coat and wander outside to witness the northern lights. Other times, they woke me from a sound sleep. I would steal from the bed to watch green ribbons of light dip and turn. One night the Sky Dancers gathered at an apex overhead and moved in long streaks toward the earth. They wore every colour, as if some sky god had clapped a bowl over the world and told them to frolic in lines of light from apex to horizon.

After decades as a city woman, the pleasures of country life still surprised me. I loved the quiet, the turning seasons, the wildlife, and the animals in our care. It was not a life I would have chosen, yet each day held its own story, its own grace. Nights beneath the northern lights, mornings jeweled with dew, the hush of falling snow—these are the moments I will always carry, awed and forever changed by a wildness I never sought.

Chapter Sixty-Six
BEARS IN MY BACK YARD

The first bear appeared at a comfortable distance. In spring, when snow melted and lush green growth appeared on the hill beyond our fields, a black bear wandered in. He plopped down on his bum and feasted on the ursine version of a spring tonic, swiping pawfuls of greens into his mouth. We watched at our leisure, appreciating the creature without fear.

Then came the year of the predator bear. We were hosting a music festival at the ranch. Performers arrived early, sleeping in tents dotted around the property. Around six in the morning, one of them shouted beneath our bedroom window. "Bear!" We pulled on clothes and raced outside. Richard grabbed the shotgun as we ran toward the small pasture where our Shetland sheep were grazing. A young bear, likely three years old, was in hot pursuit of breakfast. Luckily, the Shetlands were quicker than the bear, keeping out of its path long enough for Richard to take a shot.

Fortunately for the bear, he missed. He climbed over the fence and was out of sight before Richard could reload. Other homes likely offered easier pickings. That particular bear never returned.

Then came the summer the bears stayed. A black bear mother

and her three cinnamon cubs settled in the forested area on the southeast corner of the ranch. The cubs romped and tumbled, played through the long northern summer days, always under Momma's watchful eye. She, in turn, kept an eye on us.

By then, we had Akbash guardian dogs, so we did not worry about predation. Dogs and bears gave each other a wide berth, which allowed our sheep to graze in peace.

One day, walking through the tall trees in the southeast corner, Richard said, "Look up!" High overhead, we saw the fuzzy brown rump of a bear cub perched in one of the largest firs. We searched for Momma Bear but couldn't see her through the thick limbs. We knew we needed to move on quickly. Coming between a mother and her cubs is like waving a red flag.

For once, Phantom, our usual barking doofus, stayed silent. Perhaps he missed the overhead cue. We walked on, alert, Phantom close at heel. That day we never saw the mother, though she was undoubtedly nearby.

That summer, the bear family watched from a respectful distance as we worked the fields. The cubs tumbled, foraged, and scuffled under Momma's eye. Their presence felt like a gift—a glimpse into wild lives quietly unfolding alongside our own. In the hush of the forest, I realized that life in the country was full of surprises, from the smallest heartbeat of a newborn lamb to the wary watch of a bear overhead.

Just down the road, another kind of wildness waited, one measured in hooves, horns, and human courage: the rodeo.

Chapter Sixty-Seven
CHEERING FOR THE BULL

During our years on a small ranch in the heart of British Columbia's Cariboo country, I attended many rodeos. I admired the skill, speed, and courage on display, yet I was troubled to see animals goaded into fear or anger, and I worried when they were injured.

Bull riding was my main guilty pleasure because it was such an uneven match. The advantage was always on the bull's side. I figured the riders were crazy to climb onto the backs of these powerful animals and provoke them to buck. We had met retired bull riders at cowboy poetry festivals. They were a proud, crippled lot: bad backs, missing fingers, wrenched joints, massive belt buckles, and a swaggering walk.

During the "Animal Athletes" tour at the Williams Lake Stampede, I learned how much the gate and its orientation could affect a bull's bucking style. Even the flank strap's tightness mattered. Less bucking meant less money for the cowboy, so smart riders treated the gate crew with respect.

Between rides, the bulls lounged with their pals, enjoying a peaceful life. Each was worth $25,000 or more—top-tier bucking bulls are as rare as world-class human athletes. The riders were a different story. Being tossed around on the back of a bull left bodies wrecked: backs wrenched, fingers snapped off, shoulders dislocated. At cowboy poetry festivals, they wore more than the massive belt buckles they won. They wore their scars and old injuries like badges of honour.

One July afternoon at the Williams Lake Stampede, a bull rider's hand got caught in the rope. He came off the bull. His hand did not. Until the outriders could free him, he was tossed around like a rag doll. It was terrifying—and awe-inspiring—to watch.

"The Bull Rider's Wife" grew out of that mix of admiration and concern. It was my way of honouring both the courage of the riders and the quiet strength of those waiting at home. I knew I had captured it when a bull rider's wife posted the poem online, wanting people to understand her life.

The Bull Rider's Wife[1]

The bull rider's wife is waiting tonight
For her bull riding man to come home
If it's his voice she hears when she picks up the phone,
She won't dance the last waltz alone.

"I pulled a good wild bull today,"
He tells her when he rings.
"I'll walk away with an 85,
Hell, a 95," he sings.
"We'll start our herd, buy land and tools.
I'll plant your apple tree.
A few more wins, and I'll head home
You save that waltz for me."

"Sam's mad," she says, "If you're not back
He'll fire you in a streak.
A storm came through and flattened hay

You should have baled last week.
Because you didn't fix the fence
A bull bred Sam's prize cow.
The boar ate Sam's wife's champion hens.
She's on the warpath now."

"I'm riding on Bodacious, Hon,"
The rider tells his wife,
"Had a bonus round in Calgary;
It's all part of the life.
So please don't fret; I'm earning cash
Enough for any bill.
Before too long, I promise you,
Our bank account will fill."

"Bodacious smashed Tuff Hedeman's face,"
She says and shakes her head.
"The horn that killed Lane Frost
Just might have skewered you instead.
I'd rather have you herding sheep
Than sitting on a bull
Till whistles blow or flags drop down.
Days here are just as full."

"The clowns and barrelman are tops,"
The cowboy says to her.
"And Tuff's back riding good as new.
Don't be in such a stir.
A few more rides, and I'll be home.
Tell Sam I'll fix that fence
And bring the hay in double-quick.
Sweet girl, have confidence."

"You call me 'fore you ride," she says,

"And call me when you're done.
It's not the money that I want.
I don't care if you've won."
She bites her lip and doesn't talk
About the dream she's had
Of broken bones and trampled clowns
And kids without a dad.

He says goodbye and clears his mind
And walks out to the chute.
He climbs the fence and glances
At the clowns in hot pursuit.
A bull has tossed his rider,
And the cowboy can't pull loose.
He's flapping like a rag doll
While the bull pours on the juice.

She listens to the radio,
Plays softball with the boys.
She irons her scarlet dancing dress
And puts away the toys.
She picks his favorite waltzing tune,
Puts clean sheets on the bed.
She listens for the phone to ring
And tries not to feel dread.

He lifts his leg across the bull,
Breathes deep and dips his hat.
The gate swings wide; the world shrinks to
A primeval combat.
He's riding on a hellcat.
All his senses are alert.
He'll stay the eight and win the purse
Or wind up eating dirt.

She glances up; it's 4:13.
He's off Bodacious now.
He's waving to the cheering crowd
Or broken on the ground.
She listens for the phone to ring
And stiffens when she hears.
The rider's wife knows every call
Might be the one she fears.

Watching these events, I often thought of our own farm and the wildness we lived with daily—not with cheering crowds, but with quiet respect for creatures both domestic and untamed.

---

1. The poem appeared on a CD of the same name, in the book, *Ride Around Real Slow*, and in *American Cowboy* magazine.

Chapter Sixty-Eight
ANGELS IN THE OPERATING ROOM

Menopause did not come knocking. I was in my mid-fifties, with no sign it ever would. Then a new doctor arrived in our small town. I booked a routine appointment for some female-type testing. She did not like what she saw and ordered an ultrasound. Alarmed by the results, she called me in Vancouver, where we were about to do a presentation, to say she had set up an appointment with one of Williams Lake's few specialists—a gynecological surgeon.

I can go from healthy to terminally ill in the snap of a finger or the length of a phone call. "I think your uterine irregularity is more than fibroids" left me putting on a brave front as I contemplated my funeral. While in Vancouver, I wandered into a bookstore. On a table packed with sale books, one practically leapt into my hands. *Close to the Bone* by Jean Shinoda Bolen is about dealing with terminal illness. With my mind looping versions of "Bury me not on the lone prairie," I leafed through the pages and practically ran to the register.

I devoured the book and soon found the passage my heart knew was there. Bolen compared surgery to Inanna's descent into the underworld. The Sumerian goddess made the journey to be with her sister, surrendering one by one all the symbols of her power and identity. Bolen likened that stripping away to the surgeon's healing knife. Inanna emerged whole and new. I wanted to believe I would too.

Back in Williams Lake, I learned my fibroids were a huge mass of growths, ballooning my uterus to the size of a 12-week pregnancy. Further testing showed no cancer, but the surgeon warned me I had at most a year before they would perforate my bladder, causing permanent damage.

I thought of Inanna and Bolen's image of the healing knife. Inanna survived by calling on friends and helpers. I decided to do the same. I wrote a letter to the surgeon, describing a ritual from Bolen's book. I asked the surgical team to pause before they began, to focus not on my fibroids but on the image of the healing knife and on the path we were travelling together—the path of my spiritual and physical health.

In the letter, I explained what the surgery symbolized for me as a childless woman. Instead of a gradual transition through menopause, I would suddenly join the tribe of women thrust into that new phase of life: the threshold of old age and death. In a culture that wraps older women in a cloak of invisibility, the passage felt momentous. In a society rife with stereotypes about women who are childless by choice or by circumstance, surgery was another marker of standing outside the cultural norm.

For a quarter of a century, I had tried to keep the scabs in place when thoughtless questions probed my fertility struggles. In my late twenties and early thirties, I endured every test possible to determine why I could not conceive. They were years of well-meant intrusions and bad advice:

"Have you tried [insert here: sexual position and timing, diet, psychotherapy, this exercise, that relaxation technique, visualization]?"

"A friend of mine had the same problem. As soon as they adopted their first child, she conceived."

By nature, I am private. Being called to account for my inability to have children was painful. Eventually, my gynecologist refused further testing until my husband agreed to be tested. He did so reluctantly. The answer broke us in ways we barely understood, and our marriage limped along before ending.

Decades later, facing a hysterectomy, the scars reopened. I wept. I wrote letters to a friend. I researched every possible alternative—not to become a fifty-four-year-old mother, but because the prospect of surgery unearthed long-buried sorrows.

Social circles overlap in a small town. The anesthesiologist was a jazz musician whose music I loved. The surgeon and I had been at a party together. They may have been surprised by my request to accompany me into the underworld and back, but they honoured it. And so I began the descent, held safe in a healing circle.

Afterward, every member of the surgical team checked on me. Their concern made me feel carried in their hearts. When I returned to the clinic to have the staples removed, the anesthesiologist drew a picture of the uterus they had removed. It was a three-headed creature, growing wherever it could find room. It needed to come out.

The physical scars healed in six weeks. The psychic scars took longer. There were angels in the operating room. They healed my body. The rest was up to me.

Even now, when friends travel with their grown children, or when someone assumes my childlessness was a choice, the old ache stirs. But I no longer fear it. Like Inanna, I returned carrying the marks of descent—stripped of illusion, reshaped by loss, yet more wholly myself. The healing knife did its work, and so did time.

Chapter Sixty-Nine
"HEY, I'M AN INDIAN"

Bill called us when he got the news. "Hey, I'm an Indian!" he announced, laughing into the phone.

The road to that announcement had been anything but smooth. His dogged insistence that the government acknowledge his cultural identity required years of research and wrangling with bureaucracies.

No one who looked at Bill would have doubted his heritage, but the government demanded proof. He had been officially labeled "white" for school enrollment, something his mother thought would make his life easier. It didn't.

His classmates needed no paperwork. Bill's straight black hair and coppery skin were all they required to target him. Decades later, he still bore the scars of those schoolyard years.

As an adult, Bill reclaimed his identity with pride. His people's history stretched back thousands of years before colonisers arrived. He wanted the Canadian government to acknowledge what he knew in his bones. When they refused, he navigated the dead ends, detours, and occasional clear paths of genealogy.

He came into our lives through Richard's decades of research into the British Columbia gold rush. The provincial archivist suggested Bill contact my husband, figuring he might help untangle some of the mysteries of Bill's family history. That first conversation was an eye-opener. What Bill lacked in formal education, he more than made up for in his ability to tease out clues buried deep in archives and documents.

The meagre benefits granted to Status Indians held no attraction for him. "My tribe gets five dollars and some baling twine every year," he joked. What mattered was recognition—an official acknowledgment of what he knew: he belonged to the land. He refused to accept the government's denial of his heritage.

Bill's first visit to the ranch launched a friendship that lasted until his death. His pickup would appear at the head of our driveway. Phantom would bark his usual greeting, and we would drop whatever we were doing and settle in for a good chat.

Year after year he worked to build a case so conclusive even the most recalcitrant bureaucrats could no longer deny his status. When the official papers finally came in the mail, he called at once.

"Hey, I'm an Indian!"

Not only was Bill finally, officially recognized. So were his children, whose light skin belied the heritage he had taught them to cherish. They were proud of their father and the legacy he had given them: a birthright rooted in pride and belonging.

Bill lived long enough to see his children become Status Indians. He had only a short time to enjoy that victory, but he died knowing he had given them something lasting. They could now say, with his same fierce joy, "Hey, I'm an Indian!"

Chapter Seventy
OH, WHERE HAS OUR LITTLE DOG GONE?

Pioneer Ranch was sixteen times larger than our farm on Vancouver Island. Sending the sheep out to pasture in the morning was easy. Bringing them back to the safety of the barnyard at night required a long walk and patience, waiting for the woollies to crop one more bit of grass.

So we acquired a sheep-herding dog. Mick was a small, tricolour (black, white, and tan) Border Collie, registered and trained, likely competition-ready. He had a bold heart and the breed's insatiable need for work.

Whistle Jack Mick was far cleverer than we were at herding sheep. In the morning, he eagerly kept stragglers in line as he moved them to the pastures. In the evening, he brought them back in a fraction of the time it would have taken us. But we had too few sheep to keep him fully occupied. When our attention slipped, he would send the flock back into the fields, only to round them up again. He needed other work.

Life became richer for Mick when we discovered his gift with baby animals. Sometimes weak lambs or piglets needed extra care.

We would bring them into the house and bottle-feed them. During the day, it was a sweet connection with new life. At night, it was a recipe for interrupted sleep. In my sixth decade, those feedings were exhausting.

Mick came to the rescue. During a bitterly cold spell, we realized how completely he understood the needs of the little ones. Instead of hauling baskets upstairs, we let him handle them. Circling the frightened newborn, he would curl his body around it. The lamb or piglet would settle into his furry warmth and drift back to sleep. Mick stayed attentive until we sent him outside for a break. My nights of interrupted sleep ended.

Mick was part of the family. Unlike the Akbash guardian dogs, who lived with the sheep, he spent his nights inside. Phantom never accepted this, jealous of the space Mick occupied in our hearts. The only time the two dogs worked in harmony was vole-hunting season. Mick had an unerring sense of where the voles hid; Phantom, clueless, relied on Mick to corner them before finishing off the catch. Mick never harmed the creatures; he simply loved finding them.

Voles, voracious and prolific, competed with livestock and created dangerous holes in the fields. On a day when gunfire targeted them, Mick ran in terror from the pops.

We never saw him again.

We spent hours calling neighbours, checking ditches along the highway, wandering through fields and bush. We put up posters in Williams Lake and Horsefly and contacted the local and provincial BCSPCA.

Someone in Horsefly thought he had seen Mick in the back of a pickup parked outside the café. He insisted the driver stay until we arrived to look at the dog. We raced over, but it was not Mick. Hope rose and fell repeatedly, but the sightings were never him. We worried he might have trusted a stranger and hopped into a vehicle —or worse, been struck by a car.

Even today, when I see a Border Collie with his colouring, I pause, imagining Mick circling a lamb, alert and watchful, body curled around the baby, utterly devoted.

Chapter Seventy-One
HOPING FOR MIRACLES

Angie was the gorgeous Angora goat we bought on Vancouver Island, hoping to control the blackberry bushes overtaking parts of our small farm. Her mohair fleece was lustrous. Blackberry thorns liked it too, clinging stubbornly until we freed her. We adored her, and of course, she came with us when we loaded Old Blue onto the ferry and headed for the Cariboo.

Somehow, we also acquired an Angora billy. By the time we held one of our two folk festivals at the ranch, Angie had given birth to twins. They were irresistible—covered in silky ringlets from tiny hooves to the nubs that would become horns. Angie was a natural mother, but only to one. The other, weak and struggling, was left to die.

We refused to accept her judgment. Scooping up the rejected twin, we brought him inside. We bottle-fed him, cuddled him, and improvised physical therapy. While kids normally walk within an hour, this little one took three weeks to take his first faltering steps. I became his mother, and he slept in a basket beside me. I coaxed him to stand, to step, to run. When he was ready, I even taught him to climb the stairs to my office. That's when we named him

Thumper. He thumped all over the house, following me upstairs and downstairs, wherever I went.

At around two months, we tried introducing him to his goat family. He sniffed them, then ran straight back to me. Slowly, I led him back, sitting nearby as he romped with his twin. When he finally joined his goat family fully, I slipped away.

Over the next weeks, he seemed to adjust to outdoor life, but he knew who his mother was. He would slip through the fence and follow me when I came outdoors for chores. If he was hungry, he would nibble my jeans until I brought him a bottle.

The folk festival in our backyard was heaven for him. I was outside for days, preparing for the festival, welcoming musicians and storytellers, and hosting the event. Although I repeatedly returned the kid to his family, he always slipped through the fence and ran to be with me. The sweet, mischievous little goat enchanted performers and the audience alike.

Angie had known what we refused to see. Thumper's miraculous recovery was a temporary blessing. Something inside that sweet body never worked properly. He began to hunch his little frame, as if listening to an impending train wreck inside.

We brought him back into the house and rigged up a feeder so he could eat whenever he was hungry. We cuddled and petted him, trying to coax him into health.

We failed. Our efforts to keep him alive had given Thumper four months of life and granted us weeks of laughter and love. He seemed to adapt when I returned him to his goat family, yet I wonder if the separation tugged at his heart in ways no bottle or cuddle could repair.

Chapter Seventy-Two
EAGERLY PLANTING WEEDS

Trying to grow vegetables in a microclimate that hovered between Zones 1 and 2 tested the skills I had picked up on Vancouver Island. There, we had a nearly year-round growing season. Heat-loving plants like tomatoes and basil struggled, but everything else repaid my care with stupendous generosity.

On the island, I had a small greenhouse. In winter, I pored over seed catalogs of heritage, open-pollinated vegetables, ordering enough for at least three gardens the size of our space. I carefully planted seeds in tiny pots, watered and watched over them, and insisted Richard and any visitors marvel at the first tiny leaves. Cool-weather crops went first in the garden; heat-loving varieties followed. The best specimens became seed stock for the next year. Those precious seeds, carefully dried and stored, were mostly immigrants doomed in Cariboo's colder climate. That made the calendula, tarragon, short-season radishes and lettuces, and forgiving carrots all the more sacred to me.

Shovel in hand, I faced the back-breaking work of clearing a weed-choked plot. Whatever seeds might flourish in the colder region were tucked carefully into warm soil. The growing season was short but intense. Every day, I checked for signs of life. When they appeared, I danced and blessed them.

Then a WWOOFer arrived. These young people from around the world made farming possible on both the island and in Cariboo. Wataru had spent his life in a Tokyo high-rise. He knew nothing of rural life or farming, and his English was limited, so I trained him with gestures.

In the garden, I tried to show the difference between weeds and desirable plants. "This plant BAD," I said, frowning at the ubiquitous thistles and nettles. "This plant GOOD," I smiled, pointing at the young vegetables and herbs. BAD plants went to the burn pile; GOOD plants were freed from weeds.

An hour later, I left him to his work and returned to the computer. Occasionally, I glanced out and saw tidy green rows forming. I felt grateful for his industriousness. At day's end, Wataru showered before dinner, satisfied with a job well done. I gestured my thanks toward the garden.

The next day, he resumed work with a good heart. The tidy rows expanded. I fed livestock, typed reports, prepared meals, and silently thanked him for his effort.

On the third morning, I wandered out before Wataru awoke. The garden that had held tender lettuce, radishes, herbs, potatoes, peas, calendula, leeks, and carrots now sported neat rows of thistle and nettle. The compost heap was piled high with young vegetables. My smiling "GOOD" and frowning "BAD" had been reversed in his mind.

After a hurried consultation with Richard, we reassigned Wataru to other tasks. Out of sight, I salvaged the remaining seedlings, planted them carefully, and watered them. We never returned him to the garden. The harvest was small, but we gained something just as valuable: a story of a diligent, good-natured visitor that made us laugh for years.

Wataru's Weeds, Summer '96

He came from a faraway city.
Tokyo was where that boy hailed from.
He didn't know hay
Or the flow of the day,
And he couldn't say "restaurant" or "crumb."

We kept him away from machinery
For he didn't grasp, "Danger, stay back!"
We gave him a hoe,
To the garden he'd go.
Weeds he could easily hack.

I pointed out thistles and veggies,
In language I'd use with a child.
"This plant bad; this plant good."
I thought he understood
And would tame all those weeds running wild.

He started to work with a vengeance.
Now and then I looked out and could see
How he chopped and he dug,
Giving strong weeds a tug,
Then replanting the good plants for me.

From a distance it looked like a picture,
Like a garden that's had lots of care.
All the plants in a row,
Cleared around with a hoe,
Why, this year I could enter the fair.

That night when he came in for supper
He looked tired but pleased with his work.

"Wataru," said I,
"You're a #1 guy,
And you're certainly not one to shirk."

Two days later I went to the garden
To admire what Wataru had done.
It was singular art,
Yes, a garden with heart,
Though no veggies were there, not a one.

In rows that looked set with a compass,
So perfect were they to the eye,
Not a carrot or pea,
No potatoes for me,
Only dandelions and thistles got by.

Wataru had piled all the "bad" plants
On the compost, as I'd asked him to.
Seems a Tokyo highrise
Doesn't train a guy's eyes
To know what a gardener should do.

He'd planted the healthiest specimens
And thrown all the others away.
He left nothing but weeds
Of the healthiest breeds,
As mementos of Wataru's stay.

Chapter Seventy-Three
PIG TUSKS AND CLOSE CALLS

The pink porkers who thrived on our Vancouver Island farm were ill-suited to Cariboo's colder, sunnier climate. We bought some for Pioneer Ranch in the first year, but they were prone to sunburn and stopped putting on weight when cold weather arrived in late fall.

We turned again to heritage breeds. A Tamworth boar and two sows joined our farm, and we began breeding the most beautiful pigs I had ever seen. In sunlight, their red bristles gleamed like copper. In winter, their coats thickened. Whatever the weather, they kept gaining weight.

Anyone who has raised pigs knows that one boar to two sows is hardly commercially viable, but we liked pigs. Enjoyment is profit in its own way, and they gave us plenty of that.

We named the sows Tabitha and Petunia, the boar Barney. From the moment they set hoof on the farm, they were joy on trotters. Their sweet faces seemed perpetually to smile.

Our pigs reveled in their senses. Shoving soil aside with their strong snouts to reach succulent roots, they grunted happily. Apples or carrots brought squeals of delight. Scratches behind the ears or belly rubs prompted groans of piggy pleasure. They seemed to enjoy our company as much as we enjoyed theirs, lingering in the barnyard with us.

Our pigs had a good life. We never snipped baby teeth, docked tails, cut out balls, or sawed off tusks. With acres to roam and soft hay for farrowing, they were never a danger to us or each other.

That is, they were rarely a danger.

One summer, our friends Colleen and Ian asked to hold their wedding at the ranch. We were thrilled. Ian had built the kitchen I loved. The couple had enlivened many an evening for us, and their wedding was an artistic delight and sweet memory for all who attended. It also prompted a rethink of our pig operation—but not because of the wedding itself.

The concern arose from a twelve-year-old flower girl. Fearless around animals, she and Barney quickly bonded. He followed her across the field, leaning into her small frame, soaking up the attention he adored.

We were thrilled and terrified. By then, Barney's tusks were long, strong, and sharp. He didn't need them to fend off predators —our Akbash guardian dogs handled that—but one playful toss of his head could have seriously injured the child.

After the wedding and many weeks of soul-searching, we decided to say farewell to Barney. He was put in a stall to await his final trip. Never confined before, he was not happy. He still received plenty of scratches and love in the days before departure, but none of that replaced his lost freedom.

One fateful day, Richard stepped over Barney's head to reach the feed tray. One sharp tusk connected with his right leg, gashing him from near the groin to halfway down his thigh.

Ambulance runs were old hat for me after years on the farm. Foot heavy on the gas, I sped my bleeding husband to the Williams Lake hospital. The staff were more amused than alarmed. Pig gashes were rare; tractor rollovers and saw accidents were common. The young intern on locum duty hesitated, searching through a massive medical tome for "pig bite" treatments. When he finally found the instructions—clean, close, monitor—Richard received his stitches. An hour later we were home, thankful Barney hadn't been taller.

A few days later, Barney followed a grain trail into the back of the truck and went for his last ride. Soon after, we sold the two sows and remaining piglets. The pig chapter at Pioneer Ranch had come to an end.

Chapter Seventy-Four
DEATH OF BLACK BOY

Our sweet old Black Boy died quietly, lying on a bed of hay. The day before, I had a hunch he would not be with us much longer. For the first time in his life with us, he had not come to me for a scratch. He just stood quietly in the warmth reflected from the log sheep shed. When I went to him, he accepted my attention, but when I turned back to finish the watering, he did not follow.

He had been our pal for nearly eleven years and was already a couple of years old when we got him. In that time, he had fathered dozens of lambs—strong and sturdy, often with his black fleece and magnificent horns.

Black Boy fought yearly battles with the other rams and usually came out ahead—until two years before his death, by which time we had sold most of the flock. His only competition in his last year was the young, vigorous Shetland ram, Buster. Black Boy stopped competing for the ewes and was content to stay out of the annual mating fray.

He would hear me shut the back door and follow me with his eyes. As I neared the barn, he would trot my way and stand patiently beside me, waiting for me to rub the woolless spot where his legs joined his chest. When I knelt, he lifted his nose to mine, catching my scent. After my flirtation with suicide, he began tucking his great, horned head into my shoulder, right over my heart, closing his eyes and sighing with contentment.

No matter what the day brought, I could count on his gentle patience, his unfailing friendliness. His eyes held an old intelligence, meeting mine as one sentient creature to another. I never learned his language, nor he mine, but we understood each other. His death hit me hard.

With Black Boy gone, the other sheep and our sweet guardian dog, Suli, no longer had to dodge his spiral horns to reach me. They gathered around, hoping for grain or eager for water. Once that was done, if I knelt down, they came for affection. Buster, the rusty-brown Shetland, and Suli were first—sometimes competing for my attention. The shyer ewes followed, curious why the adults liked it so much. Behind them came the lambs, sniffing my hands, unsure but intrigued.

We formed a tableau: Buster's chin on one shoulder, a ewe on the other, Suli snuggled against my side, lambs sniffing my hands. Our breathing slowed to a collective sigh of contentment.

Chapter Seventy-Five
GOOD FENCES MAKE GOOD NEIGHBOURS

When the need to pay off debts outweighed our desire to raise livestock, we sold most of our sheep and cows[1] and began leasing our cross-fenced pastures for summer grazing. Moving visiting cattle from one pasture to another was easy. They had good hay and water and required little of our time.

The day came for the cow/calf pairs to be trucked separately. We rounded up the cows the night before, ready for the owner's arrival. They were in the pasture nearest the loading area when the trucks rolled in.

While the owner and his helpers arranged the decks on their liners and decided how many cows or calves to put in each compartment, we eased the cattle from pasture to barnyard to loading pen. After a summer with us, they were used to our moving them—but once confined in the loading pen, they became anxious and began to bawl.

When the sorting started, the volume increased. On our farm, calves stayed with their mothers until they weaned themselves naturally, so this was a new experience for us. The crying seemed endless as the trucks were loaded, each compartment carefully filled to give the animals enough space to move safely—but not so much that they had no support on hills or during sudden stops. The grieving pairs could hear each other from their separate decks but could not reunite. I was close to tears the entire time. It was another reminder that my sensibilities were out of sync with livestock farming.

As we sorted one batch, the rancher noticed a cow that was not his. Her brand confirmed it: she belonged across the fence on Crown land. Several years earlier, a leaseholder had grazed a herd of cows there. Grass is always greener on the other side of the fence. One day, enough cows leaned through to knock it over. We penned the cattle, repaired the fence, and called the grateful owners.

The following summer, the whole herd broke into our pastures. Their water had run low, and they could smell our pond. Our hay was nearly ripe. The marauding cows trampled hundreds of dollars' worth before we could get them out. I called the owners again, thinking they would want to know their water was low and hoping they would help repair the fence their cows had torn down. Instead, the ranch wife burned my ears, ending with, "It's your problem!" Legally, she was right. We did the necessary repairs and had no more unwelcome visitors that summer.

When her brand appeared among the cows we were loading, we braced for another unpleasant call. We cut her out of the herd, but two calves scrambled out with her. A coordinated effort got the calves onto the truck, but the escapee could not be contained.

We waved farewell to the load of sorrowful cows and calves and repaired the barnyard fence the errant cow had broken through. She had crashed through more fences in her rush to escape, so we repaired our way along her path. We called the owners, but I doubt they ever found her.

As the cow broke fences, outsmarted our plans, and vanished into the pastures, I felt a mix of exasperation and admiration. Sometimes, freedom is a sweet reward—even if only for a moment.

Chapter Seventy-Six
OUTRAGEOUS COURAGE

With fewer animals to care for, I had more time to stretch in new directions. So when Anne Burrill, executive director of the Women's Contact Society, asked me to direct the second Williams Lake production of Eve Ensler's *The Vagina Monologues*, I agreed, albeit with trepidation.

Anne was a highly skilled community animator. We had first met when I began taking on consulting contracts, and she had sent numerous opportunities my way. She was a breath of fresh air in a community more conservative and patriarchal than I was used to. A few years into our acquaintance, I suggested that the Women's Contact Society bring *The Vagina Monologues* to Williams Lake—and they embraced the challenge.

Anne was the perfect person to lead a group of women in a production that tackled issues rarely addressed in public. The night of the show, the theatre was packed. When the last monologue ended, the audience rose to its feet, cheering wildly.

Following that success was a challenge I wasn't sure I could meet. With no prior directing experience, I doubted my ability to guide a group of brave women through such a controversial production. Anne's willingness to produce the show reassured me that we would find the cast and crew we needed. We agreed to include anyone who showed up. After some hesitant read-throughs, I began to wonder whether our optimism had been misplaced.

A cast member with dance and movement experience helped the group loosen up. We tried improvisation exercises, but mostly, we told stories. Some women knew each other well; others barely at all. What they all brought into the room was trust. As the stories drew them into the shark-infested waters of their lives, their fellow actors wove a safety net that pulled them back to solid ground.

Week after week, every woman who could attend rehearsal showed up. What initially seemed insurmountable—deciding who would read each monologue—solved itself. The women gravitated toward readings that stretched their comfort zones, resonated with them, or shed light on puzzling or disturbing experiences.

After weeks of working in a circle, always able to see each other's faces, they were ready for the stage of the community's 400-seat theatre.

Anne had assembled a talented, hard-working crew to handle lighting, costumes, and publicity. We found roles for the twenty-nine women who wanted to be on stage—some solo, some in choral readings. One woman shared a harrowing personal story; Donna Milner contributed a deeply moving poem. Their combined energy rocked the hall and brought the audience to its feet. Every woman delivered a top-notch performance.

In the weeks following the production, cast members stopped me on the street or sent emails to share its impact—how strangers had approached them, sharing their own stories. The heroines of this production were the women on stage, willing to speak honestly about experiences often hidden even from friends.

Telling their stories to each other gave the untrained cast the strength and freedom to deliver the monologues as if they were their own. What I had hoped would happen did: their participation became a story in itself.

Directing the play reminded me that courage and creativity can flourish anywhere, even in a small town often underestimated by outsiders. The women of Williams Lake taught me that honesty, trust, and shared vulnerability create a power stronger than any stage set or script.

Chapter Seventy-Seven
AND THEN THERE WERE CAMELS

Of all the animals that passed through our Cariboo farm, none caused more double takes than the camels. Drivers slowed, stopped to take photos, and wondered if they had somehow taken a wrong turn into the Sahara.

The Cariboo's staples were mining, forestry, ranching, and agriculture. As city transplants with a mixed menagerie, we already provided plenty of grist for the gossip mill. Still, what came next stirred more curiosity than anything else in our years there.

Marketing was never my strength, so when a new business approached GrassRoots—our consulting company—to develop a publicity strategy, Richard took the lead. He set his inventive mind to work on a campaign designed to be unforgettable. At its heart were camels.

The idea sprang from a quirky footnote in Cariboo history. In 1862, a group of entrepreneurs imported twenty-three camels to haul freight to the goldfields. It was a short-lived experiment. Camels' soft, gel-like feet were made for sand, not the rocky, muddy mountain trails of British Columbia. Even with leather booties, the poor Bactrians soon suffered.

Their smell didn't help. Horses would bolt to escape it, sometimes straight off steep trails. Mules were less skittish but still wary of the enormous, odorous beasts. When the wagon road was completed, the camels were no longer needed and were turned loose to fend for themselves.

Richard's marketing plan was audacious: feature camels in the company's grand opening. As a prelude, we would take them on a trek to Wells, accompanied by wagons and mules—a re-creation of the ill-fated 1862 pack train. The idea was eccentric enough to attract attention.

The search for camels began. Importing animals was off the table, which narrowed the field but made the task simple. Before long, we became hosts to two improbable guests: a dromedary gelding and a Bactrian bull. The company paid for their purchase; we provided the feed, vet care, and costs for a truck, trailer, covered wagon, and travel. After the marketing campaign, the camels would be ours—for business or whatever publicity stunts we could dream up.

I had adapted to chickens, turkeys, ducks, geese, sheep, cattle, horses, and mules. But camels? The very thought intimidated me. The old saying "the straw that broke the camel's back" took on a whole new meaning.

Chapter Seventy-Eight
MAKING A CAMEL FEEL AT HOME

To haul the camels, we needed a specially built trailer. Old Blue, our 1967 stock truck, couldn't pull that much weight, so we bought a sturdier one. Richard found a used camper for the back, giving him a place to sleep when he took the camels to fairs or special events. For the publicity trek, we also needed a covered wagon and a team of mules to pull it.

The idea of a camel trek through the Cariboo drew attention long before we even hitched a mule. Local, provincial, and even national reporters called almost daily. We were used to media interest in Pioneer Ranch—Richard's years as a journalist meant he knew how to make a story travel—but the camels were a whole new species of headline.

In late April 2002, the first of our surprising guests arrived. Dundas, a Bactrian bull, came from a petting farm on Vancouver Island. His owner insisted on delivering him personally, wanting to help his big friend settle in.

The truck and trailer rumbled down our driveway around eight in the morning. Richard and Dave, our hired hand, were finalizing preparations while our animals lined up along the fence to see what was coming. Mules on one side, cows on the other—each pair of eyes fixed on the strange, two-humped arrival.

Dundas stepped off the trailer, his humps wobbling, his shaggy coat shedding in clumps. He looked bedraggled but magnificent, a creature from another world blinking in the Cariboo spring light. I took a deep breath and crossed my fingers.

The mules, Red and Babe, came to the fence for a closer look. One sniff was all it took. They bolted, circled the field in alarm, and then crept back to stare again, ears flicking.

We led Dundas into the pen we'd prepared for him. He stood calmly while we scratched his thick neck and sides, then turned his soft lips toward us for what felt like the world's sloppiest kisses. Red eventually lost interest, but Babe stayed rooted to the spot, wide-eyed and snorting.

After coffee, Dundas's soon-to-be-former owner went out to say goodbye. He wanted to do it alone. He'd raised Dundas from a six-month-old calf, and after six years together, the parting was clearly hard. I watched him rest his hand on Dundas's neck, his head bowed for a long moment before he turned away. Dundas seemed to understand something was ending.

When the truck pulled out, the sad camel stared after it. All day he was unsettled, gazing down the road where he'd last seen John. He swayed from side to side, over and over, as if comforting himself in grief.

Richard and Dave had gone to town to work on converting our flatbed into a camel hauler. I screwed up my courage and went down alone to check on Dundas, armed with carrots and grain. I had never been alone with an animal that weighed over a thousand kilos, and I had no idea how to read his signals. I scratched his ears, let him nuzzle me with his enormous, soft lips, and looked into those deep brown eyes fringed with impossible lashes. He accepted the carrots, the grain, and my company, but soon resumed his steady swaying. I stood with him for a long time, feeling protective and utterly out of my depth, wishing I knew how to make a sad camel feel at home.

Chapter Seventy-Nine
WHEN A CAMEL GETS URGES

On his third day at Pioneer Ranch, Dundas was still swaying from side to side for hours at a time. He only stopped when he heard a truck approaching. Each time, he'd lift his head expectantly, then return to his weaving when the sound faded.

It looked like a bad case of camel loneliness. None of our other animals showed the slightest interest in befriending him. His only company was the occasional human who stole a few minutes from work to give him a scratch or bring feed and water. From one of his former owner's calls, we learned Dundas had lost two close friends —his human and a llama who'd been his pasture mate. No wonder he was grieving.

We were about to acquire a second camel, a dromedary named Jasper, who lived near Red Deer, Alberta. He, too, would soon face a loss, separated from his beloved companion, Larry. We hoped the two lonely bachelors might find solace in each other.

The new camel trailer was finally ready. Richard hooked it to the truck, and we drove to Red Deer to collect Jasper. The gelding was nearly as large as Dundas but carried himself with a cheery confidence that was immediately endearing.

When we brought him home, we put the two camels in adjoining pens so they could get used to each other without having to compete for space or rank. They wandered to the shared fence, sniffed curiously, caught a whiff of each other's distinct scent, and drifted apart again—Dundas to his rhythmic swaying, Jasper to exploring his new surroundings.

The mules watched in nervous fascination, edging close, bolting away, then creeping forward again to stare. The trek was six weeks away; by then, we hoped the camels would be friends and the mules would have stopped acting as if they'd seen ghosts.

The day after Jasper arrived, I found Dundas hunched down, lurching forward and backward in a way that alarmed me. Certain he was having a seizure, I ran for Richard. He hurried to the pen, took one look, and burst out laughing.

"He's not sick," he said. "He's masturbating on the water barrel."

He was right. Our bull camel had a habit that shocked the uninitiated. Fortunately, the water barrel was large enough to withstand his attentions, but visitors who witnessed the performance were rarely prepared for it. We never knew when the mood would strike him, so some guests got a memorable demonstration of camel passion.

A few days later, we brought Dundas and Jasper together in a training pen. Jasper immediately made it clear he planned to be the dominant one, and Dundas seemed content to take second place. After a few more days across the fence, they settled into a relaxed companionship.

The next challenge was to prepare them for the upcoming trek. Our hired hand, Dave, discovered that neither camel had much enthusiasm for new experiences—and both outweighed him several times over. Jasper resisted the lead rope and daily walks, while Dundas delivered his own form of protest. He loved treats but quickly learned the difference between a reward and a bribe. If he was offered a carrot for good behaviour, he accepted it graciously. If it was a bribe to coax him into something he didn't want to do, he spat into the outstretched hand.

I was the lucky one. My role was all feeding and friendliness, with no demands attached. Much to the dismay of the sheep, Dundas and Jasper began following me on my daily rounds. My woolly friends couldn't compete for apples or attention, so I'd distract the camels with their treats, then sneak a few moments with the sheep.

When the day came to truck the camels to Cottonwood House—the historic ranch where the trek would begin—the yard was abuzz. We would be setting out from there with three wagons, six horses, two mules, one dog (dear old Phantom), and a lively assortment of drivers, swampers, partners, and media folk.

The horse owners had declined our invitation to introduce their animals to the camels before departure. Remembering how horses had once plunged to their deaths rather than share a trail with the camel pack train of 1862, I couldn't help wondering just how far our liability insurance might have to stretch.

Chapter Eighty
AN UNPLANNED RODEO

Dave and Clem, a young WWOOFing volunteer from Germany, spent the night at Cottonwood House looking after our mules, Red and Babe, and the two camels. Richard and I had the luxury of a real bed at the Billy Barker Hotel in Quesnel.

By the time we returned the next morning, the camels already had fans. Dundas was in his element, basking in the attention of curious onlookers. The trek would test the more reclusive Jasper, who was friendly only when the mood struck. When he'd had enough of strange faces and unsolicited pats, he simply folded himself down for a nap.

After camping overnight, we finished packing the wagon, loaded the camels with the light packs they'd practised carrying, took deep breaths, and set out for Barkerville.

We had barely turned onto the Barkerville Highway when Dundas's pack slipped, tangling around his legs. Dundas panicked. Dave panicked.

"Cut the ropes!", Richard shouted.

"I don't have a knife!", Dave cried.

Richard tossed him his own. By then, Dundas was kicking and jumping, trying to break free.

"Release the cinch!", Richard yelled. Dave reached for it and managed to drop the pack. But Dundas's legs were already caught in the ropes, and he was thrashing wildly.

A teammate leapt from the wagon, grabbed Dundas's lead, and narrowly avoided being crushed as he worked to calm him. When it was over, everyone was safe—except for our nerves. Unfortunately, the CBC TV crew had filmed the whole chaotic episode.

As if on cue, Red began acting up. She loved to trot but hated standing still in harness while other teams moved out. She also despised horseflies. When the biters descended and the teams ahead started fading from view, Red went into rodeo mode. First she rubbed her head hard against Babe. Then she stamped, jerked, bucked—and caught a leg in the traces, crashing down.

Clem jumped from the wagon and threw himself on her to keep her from struggling. Another trekker ran up to help while Richard unhooked her. Red was drenched in sweat but unhurt. Through it all, Babe stood steady as stone—our unflappable rock of a mule.

Once everything was sorted and we were moving again, Red settled into a trot. Dundas was calm, but it was clear Dave couldn't manage both camels. Laurie, one of the crew, hopped down and took Jasper's lead rope. She stayed with him for the rest of the trip and proved such a gem that we hired her for the summer.

Sometimes the camels walked tied to the wagon. Other times, they lagged behind, preferring a slower pace. Jasper was the more challenging of the two—headstrong and independent. During rest stops, he had his own opinions about what he should or shouldn't do. Laurie's calm patience kept him manageable. Dundas, meanwhile, was unfailingly sociable. Jasper, though stubborn, soon began to associate the wagon with home. If led too far from it, he'd break away and run straight back, content when near his rolling refuge.

The first day was the hardest. Long, steep hills tested the mules' strength and our endurance. Clem shone as a header, managing brakes on descents, jumping out to straighten lines, heading horses whenever we stopped—always cheerful, always game.

Each night, Richard had arranged something special. Barkerville interpreters in full costume joined us for a barbecue the first evening. Afterward, three trekkers pulled out guitars, Richard tuned his autoharp, and I picked up my fiddle. For two hours, we sang around the fire until mosquitoes and no-see-ums drove us to our tents.

The second night we camped in Stanley, once a booming mining town with three hotels. Now a ghost of itself, it had one sagging building and a few piles of boards among the trees. I finally figured out the satellite phone just in time for Richard to take a call from CBC Vancouver. That interview led immediately to another from CBC Kelowna. It felt odd and amusing to be reliving 1862 while chatting live on modern technology.

The route took us over hills on the first and third days, with easier terrain in between. Jasper matched the mules' ten-kilometre-an-hour pace with ease. Dundas was slower, and a small cut on his toe widened and bled on the first day. We trailered him for the next two legs so it could heal, saving him for the grand entrance into Barkerville.

The plan to include horse-drawn wagons had seemed good in theory, but the horse people and camel people never quite meshed. We had warned that horses might panic at the sight or scent of camels and invited riders to visit Pioneer Ranch for a gentle introduction. None took us up on it.

By the end of the second day, we were hot, dusty, and longing for shade. The horse teams, leading the way, claimed the best camp spot in the trees by the river. Our camels were left with a choice between hilly brush in full sun or a patch of shade in a bog. They chose the mud.

Despite the challenges, the trek was unforgettable. On the final morning, Richard arranged for the camels to lead the procession so they could arrive in Wells and then Barkerville ahead of the horses.

Dundas, healed and regal, strode proudly through the dust, with Jasper close behind and the wagons rumbling after. The townspeople waved and laughed, enchanted by the unlikely parade that had wound its way through the mountains.

The crowds were smaller than we'd hoped, but the media loved the story. Wells and Barkerville got their burst of publicity—and we came home dusty, tired, and quietly proud, knowing we'd pulled off something as improbable as it was unforgettable.

Chapter Eighty-One
HIGH HOPES IN WELLS

As interest in the camels continued to swirl, Richard began dreaming of a shop and theatre in Wells. His years as a historical interpreter in Barkerville and his deep knowledge of the gold rush made him a natural for such a venture.

We called it *Camel Crossing*—a shop devoted to two intertwined threads from the Cariboo gold rush: camels and drovers. The camels were an imaginative but short-lived experiment; the drovers were the men who drove cattle north to feed the miners. Both had left their marks on the region, and together they gave shape to our new enterprise.

The tiny town of Wells, just seven kilometres from Barkerville, pinned its hopes on the project we had helped publicize. Optimism was contagious. A few new shops opened; others were freshly painted. We bought a modest storefront on the main road, with a small apartment tucked behind and a workshop to one side. A corral in back would house the camels—the irresistible attraction we hoped would draw people in. The workshop would become a concert hall for travelling musicians and special events.

Richard threw himself into the work with unflagging enthusiasm. I added renovating, cleaning, buying, and bookkeeping to days already crowded with consulting contracts and ranch duties. It meant another major withdrawal from my savings, but I told myself the investment might steady both our finances and our fraying marriage. If the shop and theatre found even modest success, I hoped we could sell the ranch, move to Wells, and build a quieter life around writing, publishing, and performing—the things we had always done best together.

While Richard measured, sawed, and painted, I looked after the animals and my clients. In the evenings, we scoured eBay for camel- and cowboy-themed gifts. We ordered books, CDs, knick-knacks, and edible treats to fill the shelves. Laurie worked with the camels and helped prepare for the opening. On weekends, I joined Richard in Wells—painting walls, cooking meals, tending the shop, and cleaning the apartment. Sundays, I drove back to the ranch to feed animals, do laundry, mow, make phone calls, and prepare the week's consulting work. Each day seemed to stretch a little thinner than the last.

Richard faced his own challenges. Some locals grumbled about camel droppings left on their walks. He had no time to begin the planned wagon rides between Wells and Barkerville. He was working long hours on renovations, the shop, the newspaper, and the animals. Even our dependable mule, Babe, voiced her discontent by learning to unlatch her gate and set herself free.

By late summer, the excitement had waned. Tourism was sluggish, the school faced closure, and the province's new plan to "devolve" heritage sites into public/private partnerships cast a long shadow over Barkerville's future—and by extension, Wells's.

By mid-September, the tourist season was over. The mules had already come home, and now the camels were returning too. I had missed them. Their corral in Wells had been small, their daily walks no match for the open fields of the ranch.

When the trailer arrived, Jasper bounded down, eager to be back. Dundas, ever the clumsy one, hesitated. He peered out, backed up, peered again. Depth and distance were never his strengths. When Richard climbed into the trailer to give him a gentle nudge, Dundas startled, leaped too far, and landed in an ungainly heap—unhurt but indignant. We couldn't help laughing, even as we rushed to check him over.

As night fell, the two great beasts wandered to the fence, watching for us. Despite the money worries that haunted my dreams, I slept soundly that night, comforted by their presence. Jasper was content; he had never enjoyed crowds or commotion. Dundas was restless, still swaying, still yearning for company. He had spent his youth in a petting zoo; no amount of attention was ever enough.

Richard soon returned to Wells, where he was now editor of the local paper. I stayed on at the ranch. My winter of the camels was about to begin—a season when two huge, unlikely friends would work their way deep into my heart, and one of them would nearly crush me.

Chapter Eighty-Two
A SAD FAREWELL

Determined to distract Dundas from his endless swaying, I tried new treats. I offered bananas (Jasper – Yay! Dundas – Phhhht!), lettuce (Jasper – Yay! Dundas – Phhhht!), kale and cabbage (Jasper – Phhhht! Dundas – Phhhht!).

Still hopeful, I pulled up piles of Canadian thistle, the stubborn weed that grew everywhere. Jasper came on the run, as he always did when food was offered. Dundas hung back. I held out a handful close enough for him to sniff but far enough that he had to stretch. The scent was enticing. He edged closer and took the offering. Delicious!

I turned back to pull more thistle. The next time I approached, Dundas hustled toward me. Ah, bliss! Thistle—spiny, prickly, luscious greens. Fuzzy heads, yum!

For the first time in days, Dundas abandoned his swaying, standing beside Jasper as the two camels blissfully munched through the pile. That night, they lay together in the backyard, a peaceful if curious sight.

Dundas had swayed all summer, except when nuzzling tourists. I had come to accept that he was a neurotic camel, dependent on humans to entertain him. But feasting on thistles with Jasper seemed to spark something in him. For hours at a time, Dundas was at peace.

In his short time at the ranch before being trucked to Wells, he had been a lonely and forlorn creature. I feared his old pen might hold bad memories, but he seemed drawn to it. He had lost weight over the summer; now, after eating his fill of the lush greens, his ribs once again disappeared beneath his tough hide.

With Jasper content and Dundas more settled, I became increasingly comfortable with the camels. No longer intimidated by their size, I enjoyed the company of my unusual pals.

Richard wanted to learn more about training camels and made plans to fly to Los Angeles to attend a camel school. He was dividing his time between the ranch and Wells, busy with preparations for the coming summer.

I looked after the remaining sheep and cows and kept up with contracts and bookkeeping. Some nights, the Spirit Dancers would awaken me in the quiet hours before dawn. I never minded. The wildness of their silent music and the brilliance of their swirling costumes made sleep impossible. They woke the coyotes, too, who taunted our guardian dog Suli with threats of what they would do to her sheep should she drop her vigilance for even a moment. The Northern Lights made night walkers of all of us—creatures drawn to an unnameable mystery.

As I prepared the gardens for winter that October, the camels hung over my shoulders, waiting for tasty handouts. Jasper was more interested in the treats; Dundas, in human company. His permanent teeth were still coming in, so at times he got testy with Jasper but never with me.

The animals calmed me. So did working in the garden. We were walking a financial precipice, and I worried about it constantly. The company of Jasper and Dundas eased the worries that clung to me like burrs.

Our remaining cows brought comfort too, even when they

wandered out of sight and refused to return to the night pasture. All I had to do was figure out where they were grazing in the bush and call, "Topsy!" The friendly Belted Galloway cow loved her treats and knew I only called when I had something to offer. She would come running across the field toward the open gate. The others would catch her excitement, kicking up their heels and bobbing their heads as they dashed for whatever Topsy was after. I thought back to the days when their eagerness would have sent me scrambling over a fence.

The camels still occasionally startled me when they came running for treats, but I learned an easy solution: I held up a rake or any kind of tool. They didn't like anything too near their faces and would screech to a halt.

That habit came in handy when I was raking leaves. They always preferred whichever pile was nearest to me. For Jasper, it was the possibility that the leaves closest to me might be the sweetest. For Dundas, who didn't eat as much—or as indiscriminately—as his dromedary pal, it was love. He simply liked being with humans. If they had been people instead of camels, Jasper would have been a dashing rogue, Dundas a gentle nerd.

Dundas's rut began that winter. The huge Bactrian would throw back his head to rub his scent glands on his hump and flip urine onto his back with his tail. With no female camels in sight, his efforts were wasted—but they made him smell like a sewer. The rank odour didn't bother Jasper, and the two remained inseparable.

I was accustomed to being followed by the camels whenever I went outside. Visitors were not. One day, the heating-oil truck came down the drive and stopped outside the kitchen window. The driver waved. I waved back. He waved again—more frantically. I pulled on boots and jacket and went out, only to find he was afraid to get out of the truck because Dundas had come running up to greet him. I stayed with the driver until he finished pumping the oil, then walked him back to his truck, contemplating the unexpected value of camels as guard animals.

In mid-January, I was alone at the ranch when Jasper stopped eating. The dromedary who loved his treats turned down even apples and carrots. He still seemed strong and alert, but a day later I found him lying alone in the pasture beside the house, his head twisted at an unnatural angle. Jasper—so full of mischief, so eager for treats, so alive—was gone.

The vet found a massive bruise in his belly, blood clots in his esophagus, and damage to his liver and kidney—signs of a pre-existing condition his previous owner couldn't have known about. What killed him was an abscess behind one of his eyes.

We were devastated—grieving not only for the animal we loved but also for our dreams of another season with the camels.

For days, Dundas wandered the property, searching for his friend. None of the sheep, cows, or mules showed the slightest interest in him. He began stopping by the house, staring through the window until I came out with a treat and some attention. Whenever I was outside, he followed me on my rounds. At night, he slept beneath the tree in the backyard, where he and Jasper had once lain together.

One night, I drove home from town to find Dundas blocking the driveway, lost in what could only be described as amorous enthusiasm. I couldn't persuade him to move away from the culvert so I could drive through the gate. He was far too preoccupied with his own passion. Fortunately, Richard was home and managed to prod him aside. I parked, grabbed a carrot from the house, and walked Dundas down to the barnyard. That night, he would have to sleep in his own space, away from the house and out of the way in case we needed to check on a pregnant ewe.

I should have paid closer attention to that warning. Two thousand pounds of amorous Bactrian had always been a source of laughter—until, in a single thundering nudge, humour and horror collided, and I realized just how overwhelming a camel's affection could be. That night Dundas claimed the space near the house for himself. No matter how I tried to lure him elsewhere with carrots or coax him with gentle scratches, he stayed, reminding me that his devotion had a scale all its own.

I stood there brushing the snow from my boots, marveling at the breadth of a creature's love and the unpredictable ways it could crash into your life. Little did I know Dundas's enthusiasm would soon find an even more unforgettable target: me.

Chapter Eighty-Three
STANDING BROAD JUMPS

With Jasper gone and Dundas on his own, I wanted to ease the loneliness of the big Bactrian. Our sheep loved being rubbed in the woolless spot where leg meets belly, so I tried the same with Dundas. He stood quietly, dreamily. If a camel could purr, he would have done so. He liked it so much that he would hustle up to me, stop with his side pressed close, forelegs planted beside my hand, head held high, waiting for his rub. He leaned into my touch, so relaxed and happy I feared he might topple over and crush me.

He became my constant companion when I was outside doing chores. Clumsy as a teenager in a growth spurt, he spooked easily, and only carrots had enough allure to coax him into something new. I used them to lure him into a fresh pasture behind the house. Accustomed to hay, he hesitated over the spring grass but soon acquired a taste for it.

His reward was a thorough combing with a wire brush. He was shedding his winter coat, and his skin itched. Careful not to pull the guard hairs, I brushed out the loose tufts. Dundas would *cush*—folding his legs and settling on his belly—so I could reach all around his immense body.

On a balmy spring day, I walked into the pasture for our daily grooming session. He cushed and sighed in contentment as the brush loosened the last of his winter fur.

So many thoughts tumbled through my mind. My consulting work was going well. Our sweet old dog, Phantom, was nearing the end of her life and needed constant comfort. The cows and ewes were birthing without problems. I had learned to love the Cariboo —its quiet rhythms, its closeness to nature—but I longed to be free of debt and able to focus on my own work. Canada had become a good fit. As a dual citizen, I could work on both sides of the border. Still, leaving the ranch meant starting over in my mid-fifties, with few resources and no clear path ahead.

I brushed and reflected for nearly an hour. When I finally realized how late it had become, I gave Dundas a final pat and headed for the pasture gate. I was finished with the grooming. Dundas was not. He leaped to his feet and began circling me, moaning with a whining groan I had never heard before.

Warning bells rang. A camel's rut lasts many months. Dundas was young, sexually mature, and loving was on his mind. I was alone. The fence was at least three metres away. Dundas circled, keeping me in his sights.

Visions of being crushed by an amorous bull put wings on my feet. When his circling put him behind me—and the fence between us—I leapt. My standing broad jump cleared the barrier without a touch. A videographer should have been there; surely not many untrained mid-fifty-year-olds can make such a leap.

A disappointed camel hung his head over the fence as I headed back to the house. It was the last time I brushed him from his side of the fence. Our daily grooming continued, but now from opposite sides of the barrier.

I had made the decision. This would be my last summer as a reluctant farmer. I felt alone and terrified, but when I needed it most, help came my way. And as I walked back to the house, I realized it was also time to gather the pieces of my life and figure out where I truly belonged.

Chapter Eighty-Four
FINDING MY SEALSKIN

As I packed up the life we had built together, I found myself tracing its arc—how two dreamers had set out to create something extraordinary, and how far those dreams had carried us. Before I could step into what came next, I needed to look back at where it all began.

From our very first meeting, Richard's creativity sparkled like sunlight on water. Listening to him read his stories to the audience at Pine Lodge Farm, I was struck by his gift for language and storytelling. Always brimming with ideas, he would appear at my office door, freshly showered and grinning. "I have an idea," he'd announce, and I'd know another imaginative venture was about to unfold.

He had a rare knack for problem-solving. Life on our farm on Vancouver Island presented constant challenges, but Richard could quickly assess a situation, gather the right materials, and set to work with determination. When we moved to Pioneer Ranch, where the climate was harsher and the nearest shop much farther away, his ingenuity became even more apparent. I tended to see the obstacles; he saw possibilities. Where I saw old buildings and neglect, he saw potential.

One of his first major projects was turning a dowdy house into a comfortable home. I proved surprisingly skilled with a wrecking bar but hopeless with a hammer; my fingers and the boards suffered equally. I was far more useful insulating walls, clearing decades of flies from hidden spaces, painting and papering, and keeping everyone well fed. Richard and the helpers who came to spend time with us did all the carpentry, transforming one building at a time.

The chemistry between us during performances was undeniable. Our blend of stories and songs, told from two perspectives, resonated with audiences wherever we went. Together, we started a small publishing company—Winter Quarters Press—producing two of Richard's books and two CD-and-poetry collections. We launched GrassRoots Consulting, secured contracts across the Cariboo, and hosted house concerts for travelling musicians at both the farm and the ranch.

There was much to celebrate in those years of shared work, dreams, and laughter. Yet, even with all that richness, I knew it was time to let the partnership go.

Richard was in Wells for most of the spring and summer of 2003, publishing the local newspaper, tending our Camel Crossing shop, and hosting musicians in our performance venue. Back at Pioneer Ranch, I looked after our remaining animals, mowed the large lawn, scraped and painted the house exterior, and managed consulting contracts.

In late June, I drove to Prince George to meet with Theresa Healy. We were collaborating on the federally funded HEAL project (Healthy Eating and Active Living in Northern BC). We had first met a few years earlier while co-facilitating a conference, and participants often asked how many years we had been working together. By 2003, we were close friends. I confided in her and her partner, Wendy, and without hesitation they offered to turn their upstairs into a small apartment for me.

Richard and I had been tentative with each other that summer, both aware that change was coming. I drove to Wells, and we had The Talk. He wasn't surprised. "It hasn't been good lately, has it?" he said quietly.

We could look back with pride at all we had accomplished in our twelve-year marriage, but I had always been a reluctant farmer. As I began re-plotting my life, I remembered a story I used to tell—of a selkie who drops her sealskin to dance on land. An enamoured farmer steals the skin and takes her as his wife. As long as he keeps it hidden, she lives a human life with him and the son she bears. But when she finds her skin, she returns to the sea, where her heart has always belonged.

In the four months I lived with Tess and Wendy, they helped me reclaim the sealskin I had tucked away to be a supportive wife and partner. Our weekly house meetings smoothed rough edges. We shared cooking and cleaning, laughed often, and turned ordinary evenings into small celebrations. Bit by bit, I gained the confidence to step fully into the independent life I was ready to claim.

An unexpected notice set me on a new path. Stagebridge Theatre—the oldest theatre for seniors in America—was looking for someone to manage a storytelling project. They had secured funding to train elders as storytellers in struggling schools, helping children discover the joy of reading. The job description asked for experience in education, storytelling, and community development—the very arc of my working life.

I picked up the phone and called the director. "I think you're looking for me," I said.

That call was a small act of courage, but it marked a turning point. The years of farming had taught me endurance, patience, and a deep respect for the rhythms of living things. Now, those same lessons would carry me into a new kind of harvest—of stories, of connection, of opportunity—and of a life wholly my own.

Pioneer Ranch from the Horsefly Road

Phantom sitting by a basket of wild mushrooms.

Black Boy approaching for his regular scratching.

Broody silky hens with their growing checks (left) and Bonnie with her goslings (above). All photos by Cathryn

Suli with her pups and her pups with their lamb cousins.

Mickey babysitting a Tamworth piglet born on a cold night.

Cathryn with Buster and Shetland lamb (above). Dundas appearing at our front door, wanting attention (left). Topsy with her newborn calf (below). Photo with Cathryn by Richard Wright; others by Cathryn.

EPILOGUE

The farming years remain a time of unanticipated lessons. The animals, each with their own quirks and personalities, taught me patience, respect, and the joy of unexpected friendships. The camels, with their clumsy affection and stubborn independence, left an indelible mark on my heart, reminding me that love and laughter can arrive in the most improbable forms.

Looking back, I see the arc of those years as a tapestry woven with risk, creativity, and endurance. There were moments of triumph and moments of sheer exhaustion, of fear and exhilaration, of heartbreak and delight. Yet every challenge was an invitation to adapt, to stretch, to find a way forward even when the path was uncertain.

Reclaiming my sealskin, stepping into my own life, did not mean leaving the past behind—it meant carrying its lessons, its stories, and its wonder with me. The Cariboo, with its wild rhythms, its stubborn animals, and its wide-open skies, became a classroom and a home, shaping not only what I could do but who I could be.

With my sealskin reclaimed, I carried the lessons, laughter, and love of that improbable life with me—finding home not in a place, but in the courage to be fully myself. Now, in my thirty-fifth home in seventy-nine years, I carry home on my back like a human turtle —and wherever I go, I am always at home.

ACKNOWLEDGMENTS

Thank you to the beautiful people on Vancouver Island and in British Columbia's Cariboo region who welcomed this fledgling Canadian. You offered friendship, laughter, and opportunities, filling my memories with joy, warmth, and color.

Jimmy Neil Smith, founder of NAPPS (National Association for the Preservation and Perpetuation of Storytelling), set me on this path. He loved sharing his passion for storytelling but preferred to do so from his home in Jonesborough, Tennessee. When a Canadian tourism conference asked him to speak, he called me in Seattle. "You live close. Why don't you go in my place?" I jumped at the chance. None of the adventures in this book would have happened without that consequential request.

Jeanne Hardy passed away before I could thank her properly for her enormous contribution to this book. She was my inspiration, my writing cheerleader, and one of the truest friends I have ever known.

Thank you to the people who read drafts of the book and offered thoughtful feedback: Linda Bond, Colleen Goodman, Theresa Healy, Judith Nielsen, Liz Weir, Richard Wright, and Wendy Young. Heartfelt thanks also to those who read the chapters in which they are named and agreed to be identified: Anne Burrill, Marnie Duff, Diane Dunaway, Fred Eaglesmith, Longevity John Falkner, Don Gesinger, Maureen LeBourdais (on behalf of Nancy LeBourdais), Norris Spencer, and Heidi Redl. I could not reach some of the people cited in the book, and others have died in the years since. They were all important to the story, and I used their real names because they deserve recognition for the roles they played.

SARK, Susan Ariel Rainbow Kennedy, led a fabulous writing group that that introduced me to remarkable people who remain part of my far-flung circle of creative friends. She encouraged me to believe these stories were worth telling and told me to post these words where I could see them every day: "It is absolutely compelling reading." My inner critics had a field day during the long process of writing this memoir, but thanks to SARK's image of the Inner Wise Woman, I can finally bring *The Reluctant Farmer* to completion.

I carry these gifts—friendship, guidance, and laughter—like footprints on the Cariboo trails, a camel's gentle nudge, and the lessons of land and animals, reminding me frequently that home can be carried wherever curiosity leads.

ABOUT THE AUTHOR

Cathryn Wellner loves stories—the ones we share, the ones we live, and the ones that surprise us along the way. After many years in the U.S. and Canada, she now calls South Australia home. Her memoir of a decade as a reluctant farmer reflects her lifelong belief that stories illuminate the gifts hidden in even the most unlikely chapters of life.

Photo by Carla Holm

ALSO BY CATHRYN WELLNER

Roscoe & Me: Love Letter to a Crow

Hope Wins

Feisty Aging

Photo Books

Kelowna's Waterfront

In the Hug of Hills

Children's Series

Millie's Feathered Foster Family

Turkey Baby and the Hungry Hawk

Turkey Baby Finds Her Magic

Small Scale Stories Series

That Tree Talked to Me

Parts of Me Are Still Amazing

The Disappearing Pumpkin Choir

In the Shelter of Each Other

Your Task Is to Be Admired

In the Country of Plastic

I'll Tell You a Story

Excited and Kind of Scared

In Love with the Planet

On the Small Side

www.ingramcontent.com/pod-product-compliance
Lightning Source LLC
Chambersburg PA
CBHW020522080526
44583CB00013B/695